ECONOMIC REFORMS
IN THE PEOPLE'S REPUBLIC OF CHINA
SINCE 1979

ECONOMIC REFORMS IN THE PEOPLE'S REPUBLIC OF CHINA SINCE 1979

A Bibliography of Articles and Publications in English-Language Magazines and Newspapers

by
MARLIS SCHMIDT

*Under the Guidance of
Professor George A. Thoma, Jr.
Center for Business and Economics,
Elmhurst College
and Consultant Professor,
Shanghai University of Finance and Economics*

LOCUST HILL PRESS
West Cornwall, CT
1987

© 1987 Marlis Schmidt
All rights reserved

Library of Congress Cataloging-in-Publication Data
Schmidt, Marlis, 1945-
Economic reforms in the People's Republic of
China since 1979.
Includes index.
1. China—Economic policy—1976- —Periodicals—
Bibliography. 2. China—Economic conditions—1976- —
Periodicals—Bibliography. I. Title.
Z7165.C6S35   1987   016.338951   87-2817
[HC427.92]
ISBN 0-933951-10-8- (alk. paper : lib bdg.)

Printed on acid-free, 250-year-life paper
Manufactured in the United States of America

CONTENTS

Foreword ix

Introduction xv

[000] General Economics 3
    [010] General Economics and Teaching of Economics .. 3
    [020] General Economic Theory........................... 3
    [040] Economic History 10
    [050] Present Economic System 10
        [051] Economic Policies............................ 10
        [052] Economic Conditions 14
        [053] Economic Reforms......................... 18

[100] Economic Growth, Development, Planning, Fluctuation ... 27
    [110] Economic Growth, Development, and Planning .. 27
        [111] Economic Growth............................ 27
        [112] Economic Development..................... 28
        [113] Economic Planning 29
    [120] Economic Fluctuations, Forecasting, Stabilizing, Inflation.................................. 32

[200] Quantitative Economic Methods and Data 35
    [210] Econometric, Statistical, and Mathematical Methods, Models 35
    [220] Economic and Social Statistical Data and Analysis 35
        [221] National Income Accounting 36
        [222] Input-Output 36
        [223] Prices.. 37
        [224] National Wealth and Balance Sheets 38
        [225] Social Indicators and Social Accounts 38
        [226] Productivity and Growth: Theory and Data 39

[300]	Domestic Monetary and Fiscal Theory and Institutions		41
	[310]	Domestic Monetary and Financial Theory and Institutions	41
		[311] Domestic Monetary and Financial Theory and Policy	42
		[312] Commercial Banking	43
		[313] Capital Markets	43
		[314] Credit to Business, Consumer, Etc.	44
	[320]	Fiscal Theory and Policy; Public Finance	44
[400]	International Economics		47
	[410]	International Trade Theory	47
	[420]	Trade Relations, Commercial Policy, Integration	49
		[421] Trade Relations	49
		[422] Commercial Policy	55
		[423] International Economic Integration	58
		[424] Trade Relations with the U.S.	59
		[425] Trade Relations with Other Countries	67
	[430]	Balance of Payments, International Finance	76
		[431] Balance of Payments; Mechanisms of Adjustment; Exchange Rates	76
		[432] International Monetary Arrangements	77
	[440]	International Investment and Foreign Aid	81
		[441] International Investment and Capital Market	81
		[442] International Business	86
		[443] International Aid	90
[500]	Administration, Business Finance, Marketing, Accounting		91
	[510]	Administration	91
		[511] Enterprise Organization and Decision Theory	91
		[512] Managerial Economics	92
		[513] Business and Public Administration	94

Contents   vii

	[514] Goals and Objectives of Firms................ 94
	[520] Business Finance and Investment 94
	[530] Marketing and Advertising........................... 95
	[540] Accounting 97

[600] Industrial Organization, Technological Change,
                    Industrial Studies........................ 99
    [610] Industrial Organization and Public Policy.......... 99
        [611] Market Structure: Industrial Organization
                    and Corporate Strategy 100
        [612] Public Policy Towards Monopoly and
                    Competition 100
        [613] Public Enterprises 100
        [614] Transportation..................................101
    [620] Technological Change, Innovation, Research
                    and Development........................102
    [630] Industry Studies 104
    [640] Economic Capacity 106

[700] Agriculture; Natural Resources 107
    [710] Agriculture...107
        [711] Agricultural Supply and Demand............108
        [712] Agricultural Situation and Outlook 109
        [713] Agricultural Policy, Domestic and
                    International 111
        [714] Agricultural Finance...........................115
        [715] Agricultural Marketing and Agribusiness..115
        [716] Agricultural Reforms, Responsibility
                    System 116
        [717] Rural Reforms 119
        [718] Rural Situation..................................120
        [719] Rural Industry 124
    [720] Natural Resources.......................................124
        [721] Natural Resources, Oil, Mining, Metals....124
        [722] Conservation and Pollution...................131
        [723] Energy 131
    [730] Economic Geography 135

[800] Manpower, Labor, Population 137
    [810] Manpower Training and Allocation, Labor Force and Supply 137
    [820] Labor Markets, Public Policy 138
        [821] Theory of Labor Markets and Leisure 138
        [822] Public Policy, Role of Government 138
        [823] Labor Mobility, National and International Migration 138
        [824] Labor Market Studies, Wages, Employment 139
        [825] Labor Productivity 140
        [826] Labor Markets: Demographic Characteristics 140
    [830] Trade Unions, Collective Bargaining, Labor-Management Relations 141
    [840] Demographic Economics 142
    [850] Human Capital 145

[900] Welfare Programs, Consumer Economics, Urban and Regional Economics 147
    [910] Welfare, Health, and Education 147
        [911] General Welfare Programs 147
        [912] Education 147
        [913] Economics of Health 151
        [914] Economics of the Elderly 151
        [915] Economics of Crime 152
        [916] Minorities and Discrimination 153
    [920] Consumer Economics, Levels and Standards of Living 153
    [930] Urban Economics 155
    [940] Regional Economics 157
    [950] Social Problems and Conditions 160

Author Index 165

FOREWORD

In 1979, China embarked on what Chinese leaders are now calling its "Second Revolution."
At the Third Plenary Session of the 11th Central Committee of the Communist Party of China, during the winter of 1978-79, it was announced that the focus of China's policy would shift away from the ideological emphasis upon "class struggle" that had characterized the preceding decade and, instead, focus on the goal of economic growth and "Socialist Modernization." It was decided then that "... in light of new historical conditions and practical experience, a number of major new economic measures must be taken to thoroughly transform the system and methods of economic management."[1] Since 1979, there has been an ongoing process of economic adjustment, experimentation and reform associated with achieving the "Four Modernizations" of agriculture, industry, defense and technology, and the general targets of quadrupling national output and raising per capita income from $253 in 1978 to $800 by the year 2000.

These evolving economic changes and reforms address various weaknesses that have been perceived in earlier attempts to organize the Chinese economy on the basis of a system of rigid central planning. In the "Decision of the Central Committee of the Communist Party of China on Reform of the Economic Structure," adopted by the Third Plenary Session of the 12th Central Committee of the Communist Party of China in 1984, it was pointed out in reference to the structure of central planning that:

> Following are the major defects of this structure: no clear distinction has been drawn between the functions of the government and those of the enterprise; barriers exist between different departments and regions; the state has exercised excessive and rigid control over enter-

prises; no adequate importance has been given to commodity production, the law of value and the regulatory role of the market; and there is absolute egalitarianism in distribution.... The enthusiasm, initiative and creativeness of enterprises and workers and staff members have, as a result, been seriously dampened and the socialist economy is bereft of much of the vitality it should possess.[2]

In general, the most important policy changes of the current period of economic reform involve: (1) decentralization of economic decision making, (2) the introduction of increased material incentives and market forces as a partial substitute for central planning, and (3) an "Open Door" policy toward foreign trade and investment. These general aspects of reform have been and continue to be introduced in an incremental manner but already have had significant effects on both the structure and performance of the agricultural, industrial and foreign sectors.

To date, the most dramatic and far-reaching changes in China's economy have taken place in the rural, agricultural sector. In this sector, the organization of production has shifted from the highly collectivized structure of the "People's Agricultural Commune" to the "Responsibility System."

Initially, the responsibility system took the form of attempting to link remuneration to productive effort by establishing labor or output contracts between production teams and work groups of 5-10 families. In this initial form, remuneration continued to be based on "work points" earned by time spent at work. The production team continued to be responsible for administration and management of work groups. By 1980, 80% of China's rural production teams had adopted a form of the "Work Group Responsibility System." In 1980, the system of responsibility was extended to apply to contracts with individual households as well as work groups. However, after 1981, the system of economic responsibility increasingly moved in the direction of the "Individual Household Responsibility System." The "Individual Household Responsibility System" effectively de-collectivizes agriculture by assigning allotments of land, draft animals, machinery and property rights directly to individual households. Minimum output contracts are made directly between households and the state at fixed prices, and any additional land use and production is left to the discretion of individual households. Non-contract production can be sold in "free markets" or directly to other end users. It is this variant of the economic responsibility system in agriculture that has become dominant. By 1983, the "Individual Household Responsibility System" was adopted by 95% of the agricultural units.

In 1984 agricultural reform entered its second stage with an increased introduction of market forces into the production and distribution of agricultural commodities. Aspects of the second stage of reform were set out by Premier Zhao Ziyang in a report delivered in March 1985 at the Third Session of the Sixth National People's Congress. In the area of agriculture, this stage of reform would attempt to

> rationally readjust purchasing and marketing prices of grain ... in rural areas and introduce the practice of state purchase according to contract. Price controls will gradually be relaxed for other farm and sideline products by subjecting them to market regulation....[3]

Moreover,

> To serve the development of the Socialist Commodity Economy, we should encourage the peasants to develop, on a voluntary and mutually beneficial basis, various forms of cooperation and joint management in the processing, marketing and transport of materials and should gradually strengthen the cooperative economy in rural areas. We should actively organize commodity exchanges between town and country, set up wholesale markets on a broad scale for farm and sideline products and conscientiously improve the storage and transport systems so that more farm and sideline products will find a ready market in the cities....[4]

Economic reform in the urban/industrial sector has taken place more incrementally than reform in the agricultural sector. Initially, reform focused on a shift in emphasis among the various branches of industrial production. Subsequently, reform has been broadened to include decentralization, increased autonomy for enterprises, changes in incentives, the introduction of market forces and the encouragement of collective and individual production. In 1985, priority was given to reform of the structure of wages and prices.

Beginning in 1978 with the Third Plenary Session of the 11th Central Committee of the Communist Party of China, reforms in the industrial sector initially focused on re-adjusting various branches of industrial production. Priority was given to: (1) increased growth of light industry relative to heavy industry, (2) the technological upgrading and transformation of existing capital, (3) the expansion of consumer goods production, and (4) the growth of export industries. Emphasis was also placed on the development of infrastructure investment in energy, communications and transport.

Shortly thereafter, in addition to shifts in sectoral priorities, reform began to include changes in enterprise control and the organization of production. In October 1978, experiments to expand the decision-making power of enterprises were first introduced in six enterprises in Sichuan. The number of enterprises was increased to 100 in 1979. Important aspects of these experiments included increased flexibility to (1) produce in response to consumer demand on the condition that planned targets are fulfilled, (2) market products not purchased directly by the state, (3) retain a small percentage of profit for technical renovation, and (4) introduce bonuses related to production.

By 1980, over 6,600 enterprises throughout China were participating in such experiments. Experiments with a variety of forms of enterprise organization have continued over time. In 1980 a few enterprises began to experiment with a system in which compulsory delivery of all profits to the state was replaced with a state tax on profits, with independent accounting and the assumption of sole responsibility for profits and losses. The introduction of variants of the "Responsibility System" in industry was promoted in 1981, and by the end of that year 80% of industrial enterprises had introduced various forms of the responsibility system.

Although the system of economic responsibility in industry evolved in a variety of forms and at different rates in various regions, the basic elements of these systems involved: (1) increased enterprise discretion in output and input mix, (2) more flexibility in marketing above quota outputs, (3) individual enterprise responsibility for profits and losses, and (4) an increasing focus on profit and profit-related rewards. In the summer of 1983, the role of the profit motive was introduced throughout the industrial sector by substituting a state tax on profit (average 55%) for the system of compulsory profits delivery.

By 1985, the scope of urban/industrial reform widened to include price and wage reform. Since then, the prices of an increasing number of commodities have been fully or partially deregulated. Although wage reform has been slow, it is intended that in coming years wage reform will focus on "eliminating the current irrationalities, so that the egalitarian practice of 'everybody eating from the same big pot' in the distribution of wages will be gradually abolished and a new wage system better embodying the principle of distribution according to work will be instituted."[5]

It is anticipated that refinement and implementation of urban/industrial reform as it relates to decentralization, the system of economic responsibility, the scope of market forces and wages and prices will be complete by the end of the 1980s.

Finally, in the foreign sector, China has adopted an "Open Door" policy toward international trade and investment as part of its program of economic reform and modernization. Since 1979, a substantial increase has occurred in China's trade with the rest of the world. China's basic international trade and investment strategy is to generate foreign exchange earnings through exports to finance technology imports and to use foreign funds to finance key infrastructure projects.

Since reform began in 1979, there have been substantial gains in economic growth in most areas of the economy. Between 1977 and 1986, China's real national income increased at an average annual rate of 9%. The corresponding growth rate over the 1966-to-1976 period was 4.9%. Agricultural output grew at 9% per year during the reform period, and industrial output grew at 11% per year. China's total trade in 1985 was five times as much as it was in 1976. In addition to the dramatic organizational changes in the agricultural sector, the results of reform are increasingly seen in other areas. Of total retail sales, the share of state-owned commerce dropped from 90.3% in 1976 to 40.4% in 1985. The share of retail sales of producer collectives increased from 7.9% to 37.2%, and retail sales of private enterprise rose from 2.1% to 22.4%.

It is clear that the decade of the 1980s is one of substantial change in China. The enormity of China's population alone makes its attempts to discover how to "build socialism with Chinese characteristics" an important and fascinating area of study. The progress of reform has led to a growing interest in China's economic system and has generated an enormous literature in academic and business journals and periodicals. In this bibliography, Marlis Schmidt has taken on the formidable task of compiling and organizing many of the relevant contributions that have been made to this literature since 1979. Her efforts have been thorough and meticulous, and she has succeeded in providing a comprehensive and systematic reference resource. This bibliography should be of great utility to students of the ongoing changes in China's economic and social system.

Dr. George A. Thoma, Jr.

Notes

1. Yu Guangyuan, ed., *China's Socialist Modernization* (Beijing: Foreign Languages Press, 1984), p. 37.
2. "Decision of the Central Committee of the Communist Party of China on Reform of the Economic Structure," as published in *Beijing Review* (Oct. 29, 1984), p. iv.
3. Zhao Ziyang, "The Current Economic Situation and the Reform of the Economic Structure," Report on the Work of the Government Delivered on March 27, 1985, at the Third Session of the Sixth National People's Congress, as published in *Beijing Review* (April 22, 1985), p. viii.
4. Ibid.
5. Ibid., p. ix.

INTRODUCTION

The recent economic reforms in the People's Republic of China have caused great interest and expectation in the West. Considering its population of 1.1 billion and the important role China plays in the balance of world power, any reforms or shifts can dramatically influence international relationships and world trade.

This bibliography focuses exclusively on the economic changes that have taken place since 1979 under the leadership of Deng Xiaoping. While not complete, it attempts to list all articles that have been published in major English-language business and economic newspapers and magazines. Articles of one page or under, with some exceptions, have generally been excluded; their sheer number would have made this project quite unwieldy. Monographs, also, must await another bibliographical effort, although their somewhat less ephemeral nature than the periodical literature covered here may make them more accessible to the researcher to begin with.

The classification scheme used in this bibliography is that employed by the *Journal of Economic Literature,* and I would like to thank the American Economic Association for its kind permission to use it here. While some minor adjustments were necessary, especially in the subtopics, I have made every effort to adhere to the classification system as nearly as possible. The reader should note that the system assigns a three-digit number to each component of the outline (000-950); these numbers can be found in square brackets both in the Table of Contents and in the various chapter headings and subheadings throughout the bibliography.

The order of entry within headings is alphabetical by journal title. Where multiple citations appear for the same journal, the order is chronological, from the oldest to the most recent. Journal titles have been underlined to assist the reader in following the alphabetical format.

Mention must be made concerning the form of Chinese names both in this bibliography and in American bibliographic tools in general. While the preferred style here has been to present the family name followed by the given name with no comma separating the two, I have noted that many periodical indexes regularly list authors by family name, followed by a comma, and an initialized given name. In a number of unfortunate cases, I have even come across names inverted so that the given name is listed first, followed by a comma, with the family name reduced to an initial. A concerted effort was made to insure that listings in the Author Index were indeed by family name. In the bibliographic citations herein, however, certain stylistic variants are necessarily reflected.

This data was collected under the guidance of Professor George A. Thoma, Jr., both of the Center for Business and Economics at Elmhurst College, and Consultant Professor at the Shanghai University of Finance and Economics. I want to express my sincere gratitude for his encouragement and support during my research. It is my hope that this bibliography will be of value to both the academic and the business communities, providing easy access to valuable information without the tedious search usually required.

Marlis Schmidt
Elmhurst, Illinois

ECONOMIC REFORMS
IN THE PEOPLE'S REPUBLIC OF CHINA
SINCE 1979

GENERAL ECONOMICS
[000]

GENERAL ECONOMICS AND TEACHING OF ECONOMICS
[010]

1. Zhou Shu. "Collected Works of Chen Yun Published." Beijing Review, 29 (July 21, 1986): 30-31.

2. Schram, S.R. "Economics in Command? Ideology and Policy Since the Third Plenum, 1978-84." The China Quarterly, 99 (1984): 417-61.

3. Peebles, G. "Recent Books on China: A Review Article." Comparative Economic Systems, 27:4 (Winter 1985): 53-66.

4. Cheng-Fang, Y. "The Social Sciences in China." International Social Science Journal, 32:3 (1980): 567-69.

5. Rawski, T.G. "New Sources for Studying China's Economy." [review article] The Journal of Economic History, 43 (Dec. 1983): 997-1002.

6. Mangahas, M. "Notes on Economics in China." Philippine Economic Journal, 19:3-4 (1980): 526-33.

GENERAL ECONOMIC THEORY
[020]

7. Terrill, R. "If the Chinese Twin Ideas of Yin and Yang Did Not Exist They Would Have to Be Invented Today." Across the Board, 17 (Aug. 1980): 42-52.

8. Prybyla, J.S. "Economic Problems of Communism: A Case Study of China." *Asian Survey*, 22 (Dec. 1982): 1206-37.

9. Bachman, D. "Differing Visions of China's Post-Mao Economy: The Ideas of Chen Yun, Deng Xiaoping, and Zhao Ziyang." *Asian Survey*, 26 (March 1986): 292-321.

10. Raichur, S. "Economic 'Laws,' the Law of Value, and Chinese Socialism." *Australian Economic Papers*, 20 (Dec. 1981): 205-18.

11. Xue, M. "Addendum to China's Socialist Economy." *Beijing Review*, 24 (Dec. 7, 1981): 14-16.

12. Zhao, Z. "Present Economic Situation and the Principles for Future Economic Construction." *Beijing Review*, 24 (Dec. 21, 1981): 6-36.

13. Xue, M. "Will Small Production Lead to Capitalism?" *Beijing Review*, 25 (Jan. 18, 1982): 14-16.

14. "Agreements and Differences." *Beijing Review*, 25 (July 12, 1982): 20-22.

15. "Institute an Economic Planning System Better Suited to China's Conditions." *Beijing Review*, 25 (Oct. 11, 1982): 21-23+.

16. Quihua, Q. "Research on the World Economy in China." *The China Quarterly*, 84 (Dec. 1980): 720-26.

17. Crook, F.W. "The Baogan Daohu Incentive System: Translation and Analysis of a Model Contract." *The China Quarterly*, 102 (June 1985): 291-303.

18. Hu Ruiliang. "A Rebuttal of the 'Gang of Four's' Fabrication--The 'Duality in Socialist Production Relations.'" *Chinese Economic Studies*, 12:3 (Spring 1979): 18-36.

19. Zhang Wenxiao. "Would Money in a Socialist Economy Inevitably Breed a Bourgeoisie?" *Chinese Economic Studies*, 12:3 (Spring 1979): 37-44.

20. Xu Dixin. "Chairman Mao's Contribution to the Development of Marxism on the Question of Transforming Production Relations and Developing the Productive Forces." *Chinese Economic Studies*, 12:3 (Spring 1979): 56-86.

General Economic Theory

21. Lin Zili and You Lin. "On the Relations Between Politics and Economics." Chinese Economic Studies, 12:3 (Spring 1979): 87-108.

22. Zuo Xu. "Why Did the 'Gang of Four' Limit Chairman Mao's Directives on Theoretical Questions to the Restriction of Bourgeois Rights?" Chinese Economic Studies, 12:3 (Spring 1979): 109-13.

23. Wang Haibo; Zhou Shulian; and Wu Jinglian. "Distribution According to Labor Is Not an Economic Base for the Emergence of the Bourgeoisie." Chinese Economic Studies, 12:3 (Spring 1979): 114-33.

24. Yu Guangyuan. "The Concept of Economic Effects in the Production of Material Values Under the Condition of Socialism." Chinese Economic Studies, 12:4 (Summer 1979): 6-55.

25. Ji Chongwei and Wang Zhenzhi. "On the Question of Increasing Enterprise Earnings and Accelerating Capital Accumulation." Chinese Economic Studies, 12:4 (Summer 1979): 69-74.

26. Sun Ru. "Socialist Accumulation and Enterprise Profits." Chinese Economic Studies, 12:4 (Summer 1979): 95-106.

27. Liang Wensen. "Profit and Accumulation Are Two Categories." Chinese Economic Studies, 13:1-2 (Fall/Winter 1979/80): 5-11.

28. Hong Yuangpeng. "A Few Issues Concerning Socialist Accumulation." Chinese Economic Studies, 13:1-2 (Fall/Winter 1979/80): 12-21.

29. Yu Guangyuan. "Economic Effect and the Quality of Products." Chinese Economic Studies, 13:3 (Spring 1980): 3-19.

30. Yu Guangyuan. "The Concept of Returns on Investment and the Method for Computation When Products Have Been Stripped of Their Special Forms." Chinese Economic Studies, 13:3 (Spring 1980): 46-57.

31. Zhang Chaozun, Xiang Qiyuan and Huang Zhengi. "Socialist Ownership by the Whole People and Commodity Production." Chinese Economic Studies, 13:3 (Spring 1980): 58-68.

32. Liu Mingfu. "On the Economic Form of Socialist Economy." Chinese Economic Studies, 13:3 (Spring 1980): 69-82.

33. Weiwen, Z. "Energetically Organize the Exchange Between Industrial and Agricultural Products." Chinese Economic Studies, 13:3 (Spring 1980): 83-95.

34. Liu Guoguang and Zhao Renwei. "On the Relations Between Planning and the Market in the Socialist Economy." Chinese Economic Studies, 13:4 (Summer 1980): 3-31.

35. He Jianzhang. "Problems Involving the System of Planned Management of the Economy of Ownership by the Whole People in Our Country and the Direction of Its Reforms." Chinese Economic Studies, 13:4 (Summer 1980): 32-62.

36. Ma Jiaju. "Some Theoretical Issues Regarding the Law of Value." Chinese Economic Studies, 13:4 (Summer 1980): 63-83.

37. Dong Dasheng. "Is Not the Law of Value a Law of Commodity Economy?" Chinese Economic Studies, 13:4 (Summer 1980): 84-93.

38. Meng Lian. "Is the Utilization of the Law of Value Optional?" Chinese Economic Studies, 13:4 (Summer 1980): 94-97.

39. Ji Zhengzhi. "Problems Regarding the Formation of Socialist Planned Price." Chinese Economic Studies, 13:3 (Summer 1980): 98-115.

40. Wan Dianwu. "A Suggestion That the Term 'Net Output Value' Be Used to Replace 'Gross Output Value' as the Major Economic Index." Chinese Economic Studies, 13:4 (Summer 1980): 98-103.

41. Liu, K. "On Reforming China's Economic Management System." Chinese Economic Studies, 14:1 (Fall 1980): 38-53.

42. Lardy, N.R. and Lieberthal, K. "Chen Yun's Strategy for China's Development." Chinese Economic Studies, 15:3-4 (Spring-Summer 1982): xi-xliii.

43. Yue, W. "Production, Distribution, and Allocation of the National Economy." Chinese Economic Studies, 16:2 (Winter 1982/83): 3-19.

44. Liu, G. and Shen, L. "How to Transform Cyclical Economic Fluctuations into Smooth, Expanded Reproduction." Chinese Economic Studies, 17:1 (Fall 1983): 96-104.

45. Muqiao, X. "An Inquiry into the Problems Concerning the Reform of the Economic System." Chinese Economic Studies, 17:2 (Winter 1983/84): 3-30.

46. Yefang, S. "To Raise Production Four-fold in Two Decades Is Not Only Politically Probable but Technologically Feasible." Chinese Economic Studies, 17:2 (Winter 1983/84): 50-67.

47. Yu Guangyuan. "A Scientific Study of China's Strategy of Economic and Social Development." Chinese Economic Studies, 17:2 (Winter 1983/84): 78-87.

48. Xuezeng, L. and Shengming, Y. "Raise Economic Efficiency and Accelerate the Growth of National Income." Chinese Economic Studies, 17:2 (Winter 1983/84): 88-92.

49. Li, Z. "Again on 'Standard Output': A Further Inquiry into the Distribution Form of United Production with Remuneration Calculated According to the Standard Output." Chinese Economic Studies, 17:3 (Spring 1984): 3-17.

50. Ma, H. "Strengthen Planned Economy and Improve Planning Work." Chinese Economic Studies, 17:3 (Spring 1984): 27-32.

51. Zhang, Z. "In Stressing Economic Results Attention Must Be Paid to the Law of Value." Chinese Economic Studies, 17:3 (Spring 1984): 68-75.

52. Yue, P. "Stress the Effect of Consumption on Production." Chinese Economic Studies, 17:4 (Summer 1984): 3-9.

53. Peng, T. "A Rational Economic Structure Is the Precondition for Healthy Development of the National Economy." Chinese Economic Studies, 17:4 (Summer 1984): 10-15.

54. Dai, Y. "Methods to Appraise Economic Efficiency." Chinese Economic Studies, 17:4 (Summer 1984): 72-80.

55. Yu, G. "The Key Lies in Enhancing Economic Efficiency." Chinese Economic Studies, 17:4 (Summer 1984): 99-109.

56. Deng, H. "On the Duality of Price Subsidy and the Path to Its Reform." Chinese Economic Studies, 18:1 (Fall 1984): 3-13.

57. Ji, Z. "The Profit in Planned Prices Should Be Formulated in Accordance with a Composite Profit Index." Chinese Economic Studies, 18:1 (Fall 1984): 14-33.

58. Zuo, M. "Price Ratios Among Agricultural Products Must Be Well Adjusted." Chinese Economic Studies, 18:1 (Fall 1984): 34-43.

59. Gu, S. and Yang, Ye. "The Transformed Form of Value Under the Socialist System." Chinese Economic Studies, 18:1 (Fall 1984): 44-58.

60. Gu, S. and Yang, Y. "A Further Inquiry into Value Under the Socialist System." Chinese Economic Studies, 18:1 (Fall 1984): 59-76.

61. Lin, W. and Jia, L. "The Law of Supply and Demand and Its Role in a Socialist Economy." Chinese Economic Studies, 18:2 (Winter 1984/85): 3-22.

62. Jiang, Q. "The Basis for Socialist Production Price: A Reassessment." Chinese Economic Studies, 18:2 (Winter 1984/85): 23-38.

63. He, X. "A Preliminary Inquiry into the Theory of Service Value." Chinese Economic Studies, 18:2 (Winter 1984/85): 39-57.

64. Jia, K. "A Further Discussion of the Objective Bases for the Principle of Setting Prices According to Quality." Chinese Economic Studies, 18:2 (Winter 1984/85): 58-72.

65. Wang, W. and Hong, D. "How Do We Interpret 'Value Is the Relation of Production Cost and Utility'?" Chinese Economic Studies, 18:2 (Winter 1984/85): 73-87.

66. Zhao, L. "The Problem of Reforming the Wage System in Our Country." Chinese Economic Studies, 18:3 (Spring 1985): 35-54.

67. Ji, X. "A Discussion of the Idea That Both Interpretations of Socially Necessary Labor Time Should Be Considered in the Determination of Value." Chinese Economic Studies, 18:3 (Spring 1985): 77-91.

68. Wu, J. "Developmental Guidelines in the Early Stages of the Battle for Economic Reform and Some Questions of Macroscopic Control." Chinese Economic Studies, 19:1 (Fall 1985): 40-52.

69. Lin, Z. "Socialism and the Commodity Economy." Chinese Economic Studies, 19:1 (Fall 1985): 65-80.

70. Thaxton, R. "Peasants, Capitalism, and Revolution: On Capitalism as a Force for Liberation in Revolutionary China." Comparative Political Studies, 12 (Oct. 1979): 289-334.

71. "China Notes That Marx Is Dead." The Economist, 293 (Dec. 22, 1984): 54-56.

72. "China Closes Door to Insurance Mart." [foreign insurers barred] Journal of Commerce, 365 (Sept. 20, 1985): 1A.

73. White, L.T., III. "The Political Effects of Resource Allocations in Taiwan and Mainland China." Journal of Developing Areas, 15:1 (Oct. 1980): 43-65.

74. Chow, G.C. "A Model of Chinese National Income Determination." Journal of Political Economics, 93:4 (Aug. 1985): 782-92.

75. Ahmad, V. "The Relevance of Chinese Developmental Experience to Developing Countries." Pakistan Economic and Social Review, 17:3-4 (Autumn/Winter 1979): 90-104.

76. Sawer, M. "Soviet Discussion of the Asiatic Mode of Production." Survey, 24 (Summer 1979): 108-27.

77. "Chinese Leaders Reject Parts of Marxism to Defend Recent Economic Changes." The Wall Street Journal (Dec. 10, 1984).

78. Bennett, A. "Economic Newspaper Is Gaining Clout in China." The Wall Street Journal (Feb. 26, 1985).

79. Howard, F. "Overseas-Risk Agency Mistakes Girth for Growth." The Wall Street Journal (Sept. 19, 1985).

80. Selden, M. "The Logic--And Limits--Of Chinese Socialist Development." World Development, 11:8 (Aug. 1983): 631-37.

81. Pairault, T. "Chinese Market Mechanism: A Controversial Debate." World Development, 11 (Aug. 1983): 639-45.

82. Ishikawa, S. "China's Economic System Reform: Underlying Factors and Prospects." World Development, 11 (Aug. 1983): 647-58.

ECONOMIC HISTORY
[040]

83. Yonekawa, S. "The Development of Chinese and Japanese Business in an International Perspective." Business History Review, 56:2 (Summer 1982): 155-67.

84. Zhou, S. "Seriously Study the Strategy of Our Economic Development from a Historical Viewpoint." Chinese Economic Studies, 17:1 (Fall 1983): 85-95.

PRESENT ECONOMIC SYSTEM
[050]

Economic Policies
[051]

85. Prybyla, J.S. "Key Issues in the Chinese Economy." Asian Survey, 21 (Sept. 1981): 925-46.

86. Wilson, D. "China: One Step Backward for One Leap Forward." The Banker, 130 (Feb. 1980): 39-41+.

87. "Political Stability Is the Guarantee of Economic Readjustment." Beijing Review, 24 (Feb. 9, 1981): 20-22.

88. "New Economic Laws and Regulations." Beijing Review, 24 (Nov. 9, 1981): 6-7.

89. Hu Sheng. "Relationship Between Economics and Politics." Beijing Review, 25 (Oct. 25, 1982): 19-21.

Present Economic System

90. Zhao Ziyang. "Strategic Question on Invigorating the Economy." Beijing Review, 25 (Nov. 15, 1982): 13-20.

91. "Main Points of Premier Zhao's Government Work Report." Beijing Review, 26 (June 20, 1983): 14-18.

92. "Decision of the Central Committee of the Communist Party of China on Reform of the Economic Structure." Beijing Review, 27:suppl. (Oct. 29, 1984): i-xvii.

93. "Chinese Leaders Explain Policy Decisions." Beijing Review, 27 (Nov. 5, 1984): 6-7.

94. "Three Aspects of the Open Policy." Beijing Review, 27 (Dec. 17, 1984): 4-5.

95. "NPC Session to Promote Economic Reform." Beijing Review, 28 (March 25, 1985): 4-5.

96. "Deng Xiaoping's Speech at the CPC National Conference." Beijing Review, 28 (Sept. 30, 1985): 15-18.

97. "Chen Yun's Speech." [on economic policy] Beijing Review, 28 (Sept. 30, 1985): 18-20.

98. Tian, J. "On the Present Economic Situation and Restructuring the Economy." Beijing Review, 29 (Feb. 10, 1986): iii-xv.

99. Chen, N.-R. "Readjustment Remains Goal of China's Economic Policy." Business America, 5 (May 31, 1982): 18-21.

100. "China: Candid New Figures Have a Conservative Look." Business Week (July 9, 1979): 34-35.

101. Klatt, W. "China's New Economic Policy: A Statistical Appraisal." The China Quarterly, 80 (Dec. 1979): 716-33.

102. Oksenberg, M. "Economic Policy-Making in China: Summer 1981." The China Quarterly, 90 (June 1982): 165-94.

103. Lee, P.N.-S. "Enterprise Autonomy Policy in Post-Mao China: A Case Study of Policy-Making, 1978-83." The China Quarterly, 105 (March 1986): 45-71.

104. Nee, V. "Political and Social Bases of China's Four Modernizations." The Columbia Journal of World Business, 14 (Summer 1979): 23-32.

105. Baum, R. "Political Perspective on China's Four Modernizations." The Columbia Journal of World Business, 14 (Summer 1979): 33-36.

106. Dennis, R.D. "The Countertrade Factor in China's Modernization Plan." The Columbia Journal of World Business, 17 (Spring 1982): 65-75.

107. Phillips, C.H. "China in Transition." The Columbia Journal of World Business, 20 (1985): 53-56.

108. "Better Zhao Than Mao." The Economist, 281 (Dec. 12, 1981): 11-12.

109. "The Retreat from Marx." The Economist, 293 (Oct. 27, 1984): 17-18.

110. "The Chinese Who Still Resist a Real Leap Forward." The Economist, 294 (Jan. 5, 1985): 23-24.

111. "China-on-a-Fence." The Economist, 295 (May 25, 1985): 16-17.

112. "Deng Xiaoping Measures His Step." The Economist, 297 (Oct. 26, 1985): 77-78.

113. "China and India: Two Billion People Discover the Joys of the Market." The Economist, 297 (Dec. 21, 1985): 65-70.

114. "China: Groping for a Foothold." The Economist, 299 (April 26, 1986): 42-43.

115. Liu, M. "China Brings Its Law into Line." The Far Eastern Economic Review, 104 (April 6, 1979): 125-27.

116. Delfs, R. "Abandoning Output for Output's Sake." The Far Eastern Economic Review, 113 (Oct. 1, 1981): 60-62.

117. Delfs, R. "Socialism's New Look." The Far Eastern Economic Review, 16 (May 28-June 3, 1982): 58-60.

118. Delfs, R. "Soft-Pedaling a Massive Surplus." The Far Eastern Economic Review, 118 (Oct. 1-7, 1982): 47-49.

119. Delfs, R. "China Adopts a Three-Tier Economy with More Power in Peking: Mixed from the Centre." The Far Eastern Economic Review, 118 (Oct. 15-21, 1982): 56-58.

120. Delfs, R. "How to Evade or Oppose Peking Policies Locally." The Far Eastern Economic Review, 122 (Oct. 6, 1983): 58-60.

121. Delfs, R. and Bonavia, D. "A New Kind of Socialism." The Far Eastern Economic Review, 126 (Nov. 1, 1984): 24-25.

122. Bonavia, D. "A Failed Semantic Experiment." The Far Eastern Economic Review, 129 (July 18, 1985): 95-97.

123. Lee, M. and Bonavia, D. "Socialist Balancing Act." The Far Eastern Economic Review, 130 (Oct. 10, 1985): 36-38.

124. Delfs, R. "Chen Yun: A Chilling Speech." The Far Eastern Economic Review, 130 (Oct. 10, 1985): 39-41.

125. Delfs, R. "More Pie, Less Sky." The Far Eastern Economic Review, 130 (Oct. 10, 1985): 108-10.

126. Delfs, R. "China's Fear of Flying." The Far Eastern Economic Review, 130 (Dec. 5, 1985): 64-65.

127. Lieberthal, K. "China: The Politics Behind the New Economics." Fortune, 100 (Dec. 31, 1979): 44-47+.

128. "Economic Strategies in China Are Perplexing." Journal of Commerce and Commercial, 353 (Aug. 9, 1982): 1A.

129. Cheng, C. "Leadership Changes and Economic Policies in China." Journal of International Affairs, 32 (Fall 1978): 255-75.

130. "China in Transition." [symposium] Journal of International Affairs, 39 (Winter 1986): 1-164.

131. Mirsky, J. "Market Forces into Play." New Statesman, 108 (Oct. 26, 1984): 21-22.

132. Coplin, W.D. and O'Leary, M.K. "China: The Risks of Modernization." Planning Review, 13 (Nov. 1985): 32-37.

133. Goodman, D.S.G. "The Chinese Political Order After Mao: 'Socialist Democracy' and the Exercise of State Power." Political Studies, 33 (June 1985): 218-35.

134. Friedman, J.A. "Policy Succession in China." [review article] Problems of Communism, 34 (Jan./Feb. 1985): 83-87.

135. Volti, R. "Technology and Policy: The Dynamics and Dilemmas of Managed Change." Studies in Comparative Communism, 15 (Spring/Summer 1982): 71-94.

136. Bennett, A. "China Faced with Dilemma of Trying to Open, Tighten Economy at Same Time." The Wall Street Journal (March 29, 1985).

137. Fung, Vigor. "China Mulls Proposal to Reduce State's Role in Economic Growth." The Wall Street Journal (Sept. 5, 1985).

138. Bennett, Amanda. "Peking's Blueprint for Chinese Economy to Urge Continued Growth But Caution." The Wall Street Journal (Sept. 17, 1985).

139. McFarlane, B. "Political Economy of Class Struggle and Economic Growth in China, 1950-1982." World Development, 11:8 (Aug. 1983): 659-72.

Economic Conditions
[052]

140. McGaughey, H.W. "Wonder Out of China." Across the Board, 16 (Feb. 1979): 14-26.

141. Hidasi, G. "China's Economy in the Late 1970s and Its Development Prospects Up to the Mid 1980s." Acto Oeconomica, 23:1-2 (1979): 157-91.

142. Liu, A.P.L. "Problems in Communication in China's Modernization." Asian Survey, 22 (May 1982): 481-99.

143. Wiedemann, K.M. "China in the Vanguard of a New Socialism." Asian Survey, 26 (July 1986): 774-92.

144. Bronfenbrenner, M. "China Since Mao: A Great Leap Backward?" Atlantic Economic Journal, 12:1 (March 1984): 1-11.

145. "Marked Economic Results." *Beijing Review*, 24 (April 6, 1981): 25-27.

146. "Small Shops Coming Back." *Beijing Review*, 26 (July 11, 1983): 19-21.

147. "Individual Economy Under Socialism." *Beijing Review*, 27 (Aug. 13, 1984): 25-30.

148. "Economy Booms in First 9 Months." *Beijing Review*, 28 (Nov. 4, 1985): 7-8.

149. Lin, W. "1985, 1986 and Beyond." *Beijing Review*, 29 (Jan. 6, 1986): 4-5.

150. "Brief Look at 1985 Economy." *Beijing Review*, 29 (Jan. 13, 1986): 6-7.

151. "Optimistic But Not Over-Ambitious, Says Zhao." *Beijing Review*, 29 (March 31, 1986): 5-6.

152. "Putting Economy on Even Keel." *Beijing Review*, 29 (April 7, 1986): 5-6.

153. Verzariu, P. "Role of Countertrade in China's Economic Development." *Business America*, 3 (April 21, 1980): 14-15.

154. Thompson, D.N. "Economies of the New China." *Business Quarterly*, 43 (Autumn 1979): 36-44.

155. Keidel, A. "China: Gaining Efficiency Through Western Ways." *Business Week* (May 24, 1982): 172+.

156. Prybyla, J. "China in the 1980s." *Challenge*, 23 (May/June 1980): 4-20.

157. Ranis, G. "My Three Surprises: China, 1981." *Challenge*, 25 (March/April, 1982): 1.

158. Fureng, D. "Some Problems Concerning the Chinese Economy." *The China Quarterly*, 84 (Dec. 1980): 727-36.

159. Rae, A.E.I. "Talking Business in China." *The China Quarterly*, 90 (June 1982): 271-80.

160. Wu, J. "Current Economic Conditions and Reform of the Price System." Chinese Economic Studies, 18:3 (Spring 1985): 55-76.

161. "The Chinese Economy: Capitalism in China." Current (May 1985): 31.

162. Kraar, Louis. "Capitalism and China: China after Marx." Current (June 1985): 24.

163. Orleans, L.A. "Chinese, After All, Are Also Human." Development, 21:3 (1979): 58.

164. Shoazhi, S. "Socialism; China's Conditions; Modern Science and Technology; Democratization; Development Strategies; A Tentative Discourse on the Chinese Road to Modernization." Economic Notes, 1 (1983): 24-41.

165. Beedham, B., et al. "China in the 1980s." The Economist, 273 (Dec. 29, 1979): 17-30.

166. "Audacious Gamblers Hedge a Few Bets." The Economist, 273 (Dec. 29, 1979): 21-23.

167. "Leaky Capitalist Enclaves." The Economist, 285 (Nov. 27-Dec. 3, 1982): 87-88.

168. "Get Rich More Decorously." The Economist, 295 (April 6, 1985): 31-32.

169. Bonavia, D. "Still Searching for a Miracle." The Far Eastern Economic Review, 103 (Feb. 16, 1979): 53-54.

170. Davies, D. "Putting People in the Picture." The Far Eastern Economic Review, 105 (July 6, 1979): 11-13.

171. Sit, V. "Challenge for Hong Kong." The Far Eastern Economic Review, 107 (Feb. 1, 1980): 48-49.

172. "Accentuate the Positive, Eliminate the Negative." The Far Eastern Economic Review, 107 (March 7, 1980): 45-47.

173. Breeze, R. "Price of Modernization." The Far Eastern Economic Review, 108 (June 6, 1980): 51-52+.

174. Bonavia, D. "China '80." The Far Eastern Economic Review, 109 (Sept. 26, 1980): 42-80.

175. Bonavia, D. "Disorder Under Heaven." The Far Eastern Economic Review, 111 (Feb. 20-26, 1981): 47-49.

176. Delfs, R. "Free-Market Communism." The Far Eastern Economic Review, 126 (Oct. 25, 1984): 51-52.

177. "Economics Comes First." The Far Eastern Economic Review, 127 (Feb. 28, 1985): 96-99.

178. do Rosario, L. "That Old Sleeping Dragon Is Awakening at Last." The Far Eastern Economic Review, 127 (March 21, 1985): 65-66+.

179. "Focus: China '86." The Far Eastern Economic Review, 131 (March 20, 1986): 57-110.

180. Cheng, E. "Fair Pass-Rate in a Long Test of Faith." The Far Eastern Economic Review, 132 (May 29, 1986): 87-88.

181. Terrill, R. "China Enters the 1980s." Foreign Affairs, 58 (Spring 1980): 920-35.

182. Rowan, R. "China's Creeping Capitalism." Fortune, 104 (Dec. 28, 1981): 90-99.

183. Lardy, N.R. "China's Economic Readjustment: Recovery or Paralysis." Giornali degli Economisti e Annali di Economia, 39:7-8 (July/August 1980).

184. Clutterbuck, D. "China Looks to the Future." International Management, 35 (Sept. 1980): 16-17+.

185. Lyons, T.P. "China's Cellular Economy: A Test of the Fragmentation Hypothesis." Journal of Comparative Economics, 9:2 (June 1985): 125-44.

186. White, L.T. "Political Effects of Resource Allocations in Taiwan and Mainland China." The Journal of Developing Areas, 15 (Oct. 1980): 43-65.

187. Mirsky, J. "Cooked-Up Figures Feed No Workers." New Statesman, 110 (Oct. 4, 1985): 20-21.

188. "Storms of Change: China Reels Under the Onslaught." New York Times (Aug. 15, 1985): A23.

189. "Where China Is Booming." New York Times (Oct. 4, 1985): 30.

190. Knight, B., et al. "The Marxist World: Lure of Capitalism." U.S. News & World Report (Feb. 4, 1985): 36-42.

191. Pairault, T. "Chinese Market Mechanism: A Controversial Debate." World Development, 11:8 (Aug. 1983): 639-45.

192. Trescott, P.B. "Incentives Versus Equality: What Does China's Recent Experience Show?" World Development, 13:2 (Feb. 1985): 205-17.

193. Klatt, W. "China's Economy in the Year of the Cockerel." World Today, 37 (Sept. 1981): 348-55.

Economic Reforms
[053]

194. Knaak, R. "Economic Reform in China." ACES Bulletin, 23:2 (Sept. 1981): 1-29.

195. "Chinese Economic Reforms." [symposium] American Economic Review, Papers and Proceedings, 73 (March 1983): 319-32.

196. Reynolds, B.L. "Economic Reforms and External Imbalance in China, 1978-81." American Economic Review, 73:2 (May 1983): 325-28.

197. Gasper, D.R. "China: Striding Toward the '90s with Purpose." American Import/Export Management, 97 (Dec. 1982): 16-18+.

198. Lieberthal, K. "China's Perilous Leap into the 80's." Asia, 3 (Nov. 1980): 4-7+.

199. Prybyla, J.S. "Changes in the Chinese Economy: An Interpretation." Asian Survey, 19 (May 1979): 409-35.

200. Solinger, D.J. "Economic Reform Via Reformulation in China; Where Do Rightist Ideas Come From?" Asian Survey, 21 (Sept. 1981): 947-60.

201. Prybyla, J.S. "The Chinese Economy; Adjustment of the System or Systemic Reforms?" Asian Survey, 25 (May 1985): 553-86.

Present Economic System

202. Xu, D., et al. "China's Search for Economic Growth." *Australian Bulletin of Labour*, 10:4 (Sept. 1984): 196-210.

203. "China Moves to Counter Effects of Cultural Revolution." *Aviation Week*, 118 (June 6, 1983): 79.

204. Balassa, B. "Economic Reform in China." *Banca Nazionale del Lavoro--Quarterly Review*, 142 (Sept. 1982): 307-33.

205. Wilson, D. "Two Giants on the Move." *Banker*, 128 (Dec. 1978): 43-45+.

206. Wilson, Dick. "The Privatization of Asia." *Bankers Magazine*, 167 (Sept. 1984): 47.

207. Zhou, J. "Further Economic Readjustment: A Break with Leftist Thinking." *Beijing Review*, 24 (March 23, 1981): 23-25.

208. "Special Feature Economic Readjustment." *Beijing Review*, 24 (March 23, 1981): 26-29.

209. He, J. "Newly Emerging Economic Forms." *Beijing Review*, 24 (May 25, 1981): 15-18.

210. "No Longer an Unrealistic Target." *Beijing Review*, 25 (Nov. 8, 1982): 18-19.

211. "Reform Holds the Key to Success." *Beijing Review*, 26 (May 16, 1983): 15-20.

212. "Second Step in Economic Reform." *Beijing Review*, 27 (July 16, 1984): 9-10.

213. "Major Reform Under Way in Commerce." *Beijing Review*, 27 (Aug. 27, 1984): 10-11.

214. "Programme for Economic Structure Reform." *Beijing Review*, 27 (Oct. 29, 1984): 4-5.

215. "Session Approves Economic Reforms." *Beijing Review*, 27 (Oct. 29, 1984): 6-8.

216. "Market to Replace the Quota System." *Beijing Review*, 28 (Jan. 14, 1985): 7-8.

217. "Reforms Invigorate 1984 Economy." Beijing Review, 28 (March 11, 1985): 15-17+.

218. "Remove Obstacles to Economic Reform." Beijing Review, 28 (April 22, 1985): 4-5.

219. Zhao Ziyang. "The Current Economic Situation and the Reform of the Economic Structure." Beijing Review, 28:suppl. (April 22, 1985): iii-xv.

220. "On the Reform of Chinese Economic Structure." Beijing Review, 28 (May 20, 1985): 15-19.

221. Beng, P. "Development and Reform in Commerce." Beijing Review, 28 (July 29, 1985): 25-27.

222. "Zhao Outlines Tasks for '86 Reform." Beijing Review, 29 (Feb. 3, 1986): 6-7.

223. "Reform & Misconduct: No Direct Link." Beijing Review, 29 (Feb. 10, 1986): 26-27.

224. "Ten Major Socio-Economic Changes." Beijing Review, 29 (April 14, 1986): 19-25.

225. Phen Zhen. "Reform Conforms to Marxist Principles." Beijing Review, 29 (May 19, 1986): 14+.

226. "China Adjusts Development Strategy." Business America, 7 (Nov. 26, 1984): 2-10.

227. "China's Slow Turn Toward a Free-Market System." Business Week (May 19, 1980): 46-47+.

228. "The Greatest Leap Yet Towards a Free Market." Business Week (Nov. 5, 1984): 45.

229. Jones, D.E., et al. "Capitalism in China." Business Week (Jan. 14, 1985): 52-59.

230. Seltzer, R.J. "China's Economic Readjustment Is Aimed at More Selective Growth." Chemical & Engineering News, 59 (Dec. 21, 1981): 49-50.

231. Chihren, C. Lin. "Reinstatement of Economics in China Today." The China Quarterly, 85 (March 1981): 1-48.

232. Kueh, Y.Y. "Economic Reform in China at the Xian Level." The China Quarterly, 96 (Dec. 1983): 665-88.

233. Yeh, K.C. "Macroeconomic Changes in the Chinese Economy During the Readjustment." The China Quarterly, 100 (Dec. 1984): 691-716.

234. "The Readjustment in the Chinese Economy: Symposium." The China Quarterly, 100 (Dec. 1984): 691-865.

235. Dong, F. "China's Economy Undergoes a Sharp Change." Chinese Economic Studies, 14:1 (Fall 1980): 19-37.

236. "Why Pessimists Were Wrong on China's Free-Market Revolution." Christian Science Monitor, 77 (April 16, 1985): 3.

237. "Marx Got It Wrong: Capitalism Works." Christian Science Monitor, 77 (June 21, 1985): 14.

238. "China Reigns in Local Decision Making: Heated Economy Brings in Reassessment of Reforms." Christian Science Monitor, 77 (Aug. 6, 1985): 1.

239. Myers, H.E. "Hidden Goals in Chinese Industrialization: Lessons from Early Modernization Attempts." The Columbia Journal of World Business, 17 (Winter 1982): 74-78.

240. Brookins, C. "Perspectives on China: The New Open Door." Commodities, 7 (Nov. 1978): 44-50.

241. Bergson, A. "A Visit to China's Economic Reforms." Comparative Economic Studies, 27:2 (Summer 1985): 71-82.

242. Byrd, W. "The Shanghai Market for the Means of Production: A Case Study of Reform in China's Material Supply System." Comparative Economic Studies, 27:4 (Winter 1985): 1-29.

243. Ross, L. "Market Reform and Collective Action in China." Comparative Political Studies, 19 (July 1986): 217-32.

244. "China: The Other Foot Forward." Development, 21:2 (1979): 33-52.

245. Maitan, L. "More a Major Power ... Less a Torch or Revolution." Development, 21:2 (1979): 35-38.

246. Galting, J. "What Is Happening in China?" Development, 22:2-3 (1980): 17-22.

247. "Capitalist China." Dun's Business Month, 124 (Dec. 1984): 24-25.

248. "Anti-Red Revolutions." The Economist, 276 (Sept. 20, 1980): 11-12.

249. "Best-Laid Plans." The Economist, 286 (March 12-18, 1983): 75+.

250. "China Can Teach Russia." The Economist, 287 (June 25-July 1, 1983): 13-14.

251. "The Retreat from Marx." The Economist, 293 (Oct. 27, 1984): 17-18.

252. "Can Deng Get China to Work for Itself?" The Economist, 293 (Oct. 27, 1984): 75-76.

253. "China's Special Economic Adventure." The Economist, 296 (Sept. 14, 1985): 79-80.

254. "Reforming Reforms Is Not Revisionism." The Economist, 300 (July 26, 1986): 65-66.

255. "Land of Antiquities Looking to Modernize." Engineering News-Record, 203 (July 5, 1979): 25-28.

256. Liu, M. "China Stops to Rethink Its Development Priorities." The Far Eastern Economic Review, 103 (March 16, 1979): 106-8.

257. Bonavia, D. "Chen, The Man to Watch." The Far Eastern Economic Review, 104 (June 8, 1979): 11-13.

258. Bonavia, D. "Long March of Progress." The Far Eastern Economic Review, 105 (July 27, 1979): 23-24.

259. Liu, M. "1979: One Step Backward for One Leap Forward." The Far Eastern Economic Review, 106 (Oct. 5, 1979): 78-80.

260. Chong, P. "Invitation with Some Uncertainties." The Far Eastern Economic Review, 107 (Feb. 1, 1980): 48-49+.

261. Loong, P. "Capitalizing on New Socialism." The Far Eastern Economic Review, 109 (Aug. 15, 1980): 40+.

262. Bonavia, D. "Economic Revisionism in a Swansong Speech." The Far Eastern Economic Review, 109 (Sept. 12, 1980): 12.

263. Delfs, R. "Chen Yunis His Name, Readjustment His Game." The Far Eastern Economic Review, 113 (Sept. 25-Oct. 1, 1981): 58-59.

264. Delfs, R. "Choose Chinese Drive." The Far Eastern Economic Review, 116 (May 7-13, 1982): 54-56.

265. Bonavia, D. "What Price Socialism?" The Far Eastern Economic Review, 119 (March 10, 1983): 28-29.

266. Delfs, R. "Incentive Socialism." The Far Eastern Economic Review, 120 (April 28, 1983): 40-42+.

267. Delfs, R. "Pulling Up the Blinds." The Far Eastern Economic Review, 120 (May 12, 1983): 66.

268. Delfs, R. "China Goes to Market." The Far Eastern Economic Review, 126 (Dec. 13, 1984): 66-72+.

269. Delfs, R. "Reform Upon Reform." The Far Eastern Economic Review, 127 (March 7, 1985): 59-61.

270. "China's Changing Image." The Far Eastern Economic Review, 127 (March 21, 1985): 61-106.

271. Delfs, R. "Economic Reform: On the Road to a Second Founding." The Far Eastern Economic Review, 127 (March 21, 1985): 63-64.

272. de Wulf, L. "Economic Reform in China." Finance & Development, 22 (March 1985): 8-11.

273. Tanzer, A. "Karl Marx Must Be Spinning in His Grave." Forbes, 134 (Nov. 19, 1984): 258-59+.

274. Zagoria, D.S. "China's Quiet Revolution." Foreign Affairs, 62:4 (Spring 1984): 879-904.

275. Lieberthal, K. "Second Revolution Begins in China." Fortune, 98 (Oct. 23, 1978): 94-96.

276. Kraar, L. "China: Trying the Market Way." *Fortune*, 100 (Dec. 31, 1979): 50-54.

277. Miller, W.H. "China Flirts with Capitalism." *Industry Week*, 202 (Aug. 6, 1979): 38-41+.

278. Modic, S.J. "The New China." *Industry Week*, 224 (Jan. 7, 1985): 42-44+.

279. Ellman, M. "Economic Reform in China." *International Affairs*, 62 (Summer 1986): 423-42.

280. Ling, X.Q. "Chinese Economy Learns from Japanese Experience." *The Japan Quarterly*, 29 (Jan./March 1982), 39-48.

281. "China: New Theories for Old." *Monthly Review*, 31 (May 1979): 1-19.

282. "Greatest Leap Forward: China's New-Found Capitalism." *New York Times* (Dec. 10, 1984): A23.

283. "Still Moving Towards Market Socialism." *New York Times* (Sept. 29, 1985): F3.

284. Agres, T. "China's Efforts to Modernize, Delight, Worry Western Leaders." *Research & Development*, 27 (Feb. 1985): 119-20.

285. Chin, R.Q.P. "Modernizing China's Economy." *Rivista Internazionale di Scienze Economiche e Commerciali*, 29:7 (July 1982): 646-63.

286. "China at Large." *Round Table*, 273 (Jan. 1979): 3-11.

287. Simon, D.F. "Chinese-Style S&T Modernization." *Studies in Comparative Communism*, 17 (Summer 1984): 87-109.

288. "China's Avid Flirtation with Capitalism." *U.S. News & World Report*, 93 (July 26, 1982): 4D.

289. "China Plans a Transformation of Economy to Unpeg Prices, Reduce State Planning Role." *The Wall Street Journal* (Oct. 11, 1984).

290. "Peking Turns Sharply Down Capitalistic Road." *The Wall Street Journal* (Oct. 25, 1984).

291. "Deng Xiaoping Forces Gain Momentum in Their Attempt to Transform China." The Wall Street Journal (Dec. 13, 1984).

292. "China's Leaders Struggle to Apply Brakes, But Runaway Economic Growth Persists." The Wall Street Journal (Aug. 26, 1985).

293. Blecher, M. "The New Course in China: Summing-Up Speeches at Oxford Conference, Contemporary China Centre." World Development, 11:8 (Sept. 1982): 769.

294. McFarlane, B. "The New Course in China: Summing-Up Speeches at Oxford Conference, Contemporary China Centre, Sept. 1982." World Development, 11:8 (Aug. 1983): 767-69.

295. Nolan, P. "The New Course in China: Summing-Up Speeches at Oxford Conference, Contemporary China Centre, September 1982." World Development, 11:8 (Aug. 1983): 769-70.

296. Ishikawa, S. "China's Economic System Reform: Underlying Factors and Prospects." World Development, 11:8 (Aug. 1983): 647-58.

297. Johnson, C. "Economic Reform in China." World Today, 41 (March 1985): 55-57.

ECONOMIC GROWTH, DEVELOPMENT, PLANNING, FLUCTUATION
[100]

ECONOMIC GROWTH, DEVELOPMENT, AND PLANNING
[110]

Economic Growth
[111]

298. "Economic Growth Rate." Beijing Review, 25 (Jan. 4, 1982): 3-4.

299. Zhang Zhongji. "China's Economy: Achievements in 1983." Beijing Review, 27 (Feb. 20, 1984): 14-17+.

300. Zhu Qingfang. "Major Economic and Social Achievements." Beijing Review, 27 (Oct. 1, 1984): 16-20+; 27 (Oct. 8, 1984): 28-31.

301. "All-Around Economic Growth in 1984." Beijing Review, 27 (Dec. 24, 1984): 10-12.

302. "Economy Shows Steadier Growths." Beijing Review, 29 (July 28, 1986): 6-7.

303. Perkins, D.H. "China: Meeting Its Target for Rapid Growth." Business Week (May 5, 1980): 73-74.

304. Ishikawa, S. "China's Economic Growth Since 1949: An Assessment." The China Quarterly, 94 (June 1983): 242-81.

305. Ma, H. "Economic Adjustment and the Rate of Growth." Chinese Economic Studies, 17:1 (Fall 1983): 74-87.

306. Wu, J.; Li, J.; and Ding, N. "Hold Down the Growth Rate of the National Economy Within an Appropriate Range." Chinese Economic Studies, 19:1 (Fall 1985): 53-64.

307. "Three Economies." [economic growth of USSR, U.S., China] Christian Science Monitor, 77 (June 18, 1985): 18.

308. Malenbaum, W. "Modern Economic Growth in India and China: The Comparison Revisited, 1950-1980." Economic Development and Cultural Change, 31 (Oct. 1982): 45-84.

309. Sundrum, R.M. "Modern Economic Growth in India and China: Comment." Economic Development and Cultural Change, 34:1 (Oct. 1985): 157-60.

310. Malenbaum, W. "Modern Economic Growth in India and China: The Comparison Revisited, 1950-1980." Economic Development and Cultural Change, 34 (Oct. 1985): 157-66.

311. Malenbaum, W. "Modern Economic Growth in India and China: Reply." Economic Development and Cultural Change, 34:1 (Oct. 1985): 161-66.

312. Delfs, R. "Pace of Growth May Be Too Fast." The Far Eastern Economic Review, 118 (Oct. 1-7, 1982): 59-60.

313. Fodella, G. "China Towards a High Rate of Economic Growth: Similarities with Post-War Japan." Rivista Internazionale di Scienze Economiche e Commerciali, 30:8 (Aug. 1983): 792-97.

Economic Development
[112]

314. "Economic Readjustment: Initial Results." Beijing Review, 24 (July 27, 1981): 6-7.

315. "New Strategy for Economic Development." Beijing Review, 24 (Aug. 10, 1981): 12-17+.

316. Wang, R. "Highlights in China's Economic Construction." Beijing Review, 24 (Aug. 17, 1981): 17-21.

317. Liang, X. "China's Economic Achievements." Beijing Review, 24 (Oct. 5, 1981): 18-23.

318. Zhang Zhongji. "Economic Readjustment: Results Since 1979." Beijing Review, 25 (Aug. 30, 1982): 13-17.

319. "China Seeks Mild Growth in 1986-90." Beijing Review, 28 (Oct. 7, 1985): 6-7.

320. Wang, H. "China's Prospects for the Year 2000." Beijing Review, 28 (Nov. 4, 1985): 18-20.

321. Bennett, G. "Economy, Polity, and Reform in China." Comparative Politics, 18 (Oct. 1985): 85-100.

322. Delfs, R. "Economic Marathon." The Far Eastern Economic Review, 129 (Aug. 29, 1985): 50-51.

323. Paine, S. "Spatial Aspects of Chinese Development; Issues, Outcomes and Policies, 1949-79." Journal of Developmental Studies, 17:2 (Jan. 1981): 135-95.

324. Prybyla, J.S. "China's Economic Development: Demise of a Model." Problems of Communism, 31 (May/June 1982): 38-42.

325. Maxwell, N. and McFarlane, B., eds. "China's Changed Road to Development." [symposium] World Development, 11 (Aug. 1983): 625-770.

Economic Planning
[113]

326. Solinger, D.J. "Fifth NPC and the Process of Policy Making: Reform, Readjustment, and the Opposition." Asian Survey, 22 (Dec. 1982): 1238-75.

327. "Tasks Before Us." Beijing Review, 24 (Jan. 12, 1981): 16-19.

328. Yao, Y. "Report on the Readjustment of the 1981 National Economic Plan and State Revenue and Expenditures; Excerpts." Beijing Review, 24 (March 16, 1981): 14-20+.

329. "Communique of Fulfillment of China's 1980 National Economic Plan." [State Statistical Bureau] Beijing Review, 24 (May 11, 1981): 23-27; 24 (May 18, 1981): 17-20.

330. "Economic Questions." Beijing Review, 24 (July 27, 1981): 3-4.

331. "Communique on Fulfillment of China's 1981 National Economic Plan." Beijing Review, 25 (May 17, 1982): 15-24.

332. "National Economy: Major Targets." Beijing Review, 25 (Nov. 29, 1982): 17-19.

333. Zhao Ziyang. "Report on the Sixth Five-Year Plan." Beijing Review, 25 (Dec. 20, 1982): 10-35.

334. "Chinese-Type Modernization." Beijing Review, 26 (Jan. 3, 1983): 14-18; (Jan. 10, 1983): 13-19; (Jan. 24, 1983): 14-17+; (Jan. 31, 1983): 16-20; (Feb. 14, 1983): 14-18; (Feb. 28, 1983): 12-20; (March 14, 1983): 14-20; (March 28, 1983): 17-22; (April 18, 1983): 20-23+; (May 2, 1983): 16-19; (May 16, 1983): 15-20.

335. "Communique of Fulfillment of China's 1982 National Economic Plan." Beijing Review, 26:suppl. (May 9, 1983): ii-xii.

336. "Sixth Five-Year Plan (1981-85) of the People's Republic of China for Economic and Social Development." Beijing Review, 26:suppl. (May 23, 1983): ii-xvi; 26:suppl. (May 30, 1983): i-xvi.

337. "Report on the 1983 Plan for National Economic and Social Development." Beijing Review, 26 (July 11, 1983): i-vi.

338. "Communique on Fulfillment of China's 1983 National Economic Plan." Beijing Review, 27 (May 14, 1984): iii-xi.

339. "Report on the 1984 Economic Plan." Beijing Review, 27 (May 28, 1984): 18-20.

340. "Economic Planning Uses Law of Value." Beijing Review, 27 (Nov. 26, 1984): 4-5.

341. "Communique on Fulfillment of China's 1984 Economic and Social Development Plan." Beijing Review, 28:suppl. (March 25, 1985): i-viii.

342. "Ministers Chart Growth Plan." Beijing Review, 28 (April 8, 1985): 7-8.

343. Song Ping. "Report on the Draft 1985 Plan for National Economic and Social Development." Beijing Review, 28:suppl. (April 29, 1985): iv-viii.

344. "Sixth Five-Year Plan Succeeds." Beijing Review, 28 (Sept. 16, 1985): 14+.

345. Zhao, Z. "Explanation of the Proposal for the Seventh Five-Year Plan." Beijing Review, 28 (Oct. 7, 1985): i-v.

346. "Proposal of the Central Committee of the CCP for the 7th Five-Year Plan for National Economic and Social Development." Beijing Review, 28 (Oct. 7, 1985): vi-xxiv.

347. Geng, Y. "A Bright Outlook for Seventh Five-Year Plan." Beijing Review, 29 (March 24, 1986): 4-5.

348. "Seventh Five-Year Plan: Major Economic Indexes." Beijing Review, 29 (April 7, 1986): 27+.

349. Jian, C. "Reform Guarantees Fulfillment of New Plan." Beijing Review, 29 (April 14, 1986): 26-28.

350. "NPC Gives Go-Ahead to 5-Year Plan." Beijing Review, 29 (April 21, 1986): 5-6.

351. Zhao Ziyang. "Report on the Seventh Five-Year Plan." Beijing Review, 29:suppl. (April 21, 1986): i-xx.

352. "The Seventh Five-Year Plan of the People's Republic of China for Economic and Social Development (1986-1990)." Beijing Review, 29 (April 28, 1986): i-xxiii.

353. An Zhiguo. "Discussion on the New Five-Year Plan." Beijing Review, 29 (April 28, 1986): 4-5.

354. Song Ping. "Report on the 1986 Plan for National Economic and Social Development." Beijing Review, 29:suppl. (May 19, 1986): i-vi.

355. Chun Yun. "Planning and the Market." Beijing Review, 29 (July 21, 1986): 14-15.

356. Zhao, Z. "Report on the Sixth Five-Year Plan for National Economic and Social Development." Chinese Economic Studies, 18:4 (Summer 1985): 3-61.

357. Yu, G. "Theoretical Basis for Reform of the Planning System." Chinese Economic Studies, 19:1 (Fall 1985): 3-9.

358. "China Tries to Learn a Planner's Hardest Lesson." The Economist, 298 (Jan. 18, 1986): 29-31.

359. Bonavia, D. "Shou-Fang Syndrome in Chinese Planning." The Far Eastern Economic Review, 113 (Aug. 14-20, 1981): 56-57.

360. Delfs, R. "Laying the Foundations (Five-Year Plan)." The Far Eastern Economic Review, 118 (Dec. 10-16, 1982): 58-60.

361. Delfs, R. "No Room to Manoeuvre." The Far Eastern Economic Review, 118 (Dec. 10-16, 1982): 60.

362. Delfs, R. "The Five-Year (Non)Plan--Blueprint for the Future." The Far Eastern Economic Review, 126 (Dec. 13, 1984): 72.

363. do Rosario, L. "Planned Unplanning." The Far Eastern Economic Review, 132 (April 10, 1986): 80-81.

364. "Beijing Looks Ahead." [symposium] Problems of Communism, 28 (Sept.-Dec. 1979): 1-66.

365. "Why China Can't Plan--But Hong Kong Can." Public Interest, 72 (Summer 1983): 133-35.

366. Bennett, A. "China's Five-Year Economic Plan Seeks to Control Growth, Spending." The Wall Street Journal (Sept. 23, 1985).

ECONOMIC FLUCTUATIONS, FORECASTING, STABILIZING, INFLATION
[120]

367. Hidasi, G. "China's Economy in the Nineteen-Eighties." Acta Oeconomica, 27:1-2 (1981): 141-62.

368. "Can China Reach Its Economic Target by 2000?" Beijing Review, 25 (Oct. 4, 1982): 16-18.

Fluctuations, Forecasting, Stabilizing, Inflation 33

369. Gengmo, L. "Socialism and Inflation." Beijing Review, 25 (Nov. 1, 1982): 20-22.

370. "New Threat to China's Economy: Inflation." Business Week (Jan. 19, 1980): 38-39.

371. Xu, D. "Prospects for China's Economy in the 1980s." Chinese Economic Studies, 14:1 (Fall 1980): 6-18.

372. Chen, N.-R. and Hou, C.M. "China's Inflation, 1979-1983: Measurement and Analysis." Economic Development and Cultural Change, 34 (July 1986): 811-35.

373. Breeze, R. "Source of Inflation." The Far Eastern Economic Review, 112 (April 10-16, 1981): 78+.

374. Delfs, R. "Spectre of Inflation." The Far Eastern Economic Review, 121 (Aug. 4, 1983): 40-41.

375. Klatt, W. "China's Economy in 1985: A Review." International Affairs, 55 (Oct. 1979): 586-94.

376. Dernberger, R.F. "Prospects for the Chinese Economy." Problems of Communism, 28 (Sept.-Dec. 1979): 1-15.

377. Fodella, G. "China's Economy in the Next Twenty Years." World Today, 39 (Nov. 1983): 460-64.

QUANTITATIVE ECONOMIC METHODS AND DATA
[200]

ECONOMETRIC, STATISTICAL, AND MATHEMATICAL METHODS, MODELS
[210]

378. Travers, S.L. "Bias in Chinese Economic Statistics: The Case of the Typical Example Investigation." The China Quarterly, 91 (Sept. 1982): 478-85.

379. Yu, Q. and Gong, Z. "A Preliminary Probe into the Mathematical Model of a Systematic Analysis of the National Economic Plan." Chinese Economic Studies, 16:1 (Fall 1982): 69-87.

380. "Outline of an Econometric Model for Chinese Economic Planning." Journal of Economic Dynamics & Control, 4 (May 1982): 171-90.

ECONOMIC AND SOCIAL STATISTICAL DATA AND ANALYSIS
[220]

381. Klatt, W. "Chinese Statistics Up-Dated." The China Quarterly, 84 (Dec. 1980): 737-43.

382. Huang, H. "Problems in the Reform of Statistical Work." Chinese Economic Studies, 17:3 (Spring 1984): 61-67.

383. Luo, G. "An Analysis of the Development of China's Planned Economy and the Tortuous Course It Has Trudged: A Few Issues Concerning Socialist Planned Economy That Need to Be Clarified (Part 2)." Chinese Economic Studies, 16:2 (Winter 1982/83): 29-50.

384. Rawski, T.G. "New Sources for Studying China's Economy." Journal of Economic History, 43:4 (Dec. 1983): 997-1002.

National Income Accounting
[221]

385. "Achievements and Problems in China's 1982 National Economy." Beijing Review, 26 (April 4, 1983): 20-24.

386. Wang Bingquian. "Report on the Execution of the State Budget for 1984 and on the Draft State Budget for 1985." Beijing Review, 28:suppl. (April 29, 1985): i-iii.

387. Delfs, R. "Route Maps of the Path to 2000 and Beyond." The Far Eastern Economic Review, 129 (Aug. 29, 1985): 52-53.

388. Cheng, P.C. "Political Accounting in China: What the West Should Know." Journal of Accountancy, 149 (Jan. 1980): 76-85.

389. Kravis, I.B. "An Approximation of the Relative Real Per Capita GNP of the People's Republic of China." Journal of Comparative Economics, 5:1 (March 1981): 60-78.

390. Chow, G.C. "A Model of Chinese National Income Determination." Journal of Political Economics, 93 (Aug. 1985): 782-92.

Input-Output
[222]

391. "What Is the Aim of Socialist Production?" Beijing Review, 24 (Feb. 23, 1981): 16-20.

392. Chinn, D.L. "Basic Commodity Distribution in the People's Republic of China." The China Quarterly, 84 (Dec. 1980): 744-54.

Prices
[223]

393. "Can China's Prices Be Stabilized?" Beijing Review, 24 (May 18, 1981): 21-22.

394. "Basic Stability of Market Prices." Beijing Review, 26 (Aug. 29, 1983): 19-27.

395. "Price Reform Heralds New Economic Boom." Beijing Review, 28 (Jan. 7, 1985): 35-37.

396. Tian Jiyun. "Price System Due for Reform." Beijing Review, 28 (Jan. 28, 1985): 16-19+.

397. "Prices Go Up as Beijing Implements Reform." Beijing Review, 28 (May 20, 1985): 6-7.

398. "Market Changes to Set Prices Right." Beijing Review, 28 (May 27, 1985): 4-5.

399. "Price Reform: 6 Months Later." Beijing Review, 28 (Dec. 9, 1985): 6-7.

400. Tian, J. "China Sets to Improve Price Mechanism." Beijing Review, 29 (Jan. 27, 1986): 16-17; (Feb. 3, 1986): 15-17.

401. Liu Guoguang. "Price Reform Essential to Growth." Beijing Review, 29 (Aug. 18, 1986): 14-18.

402. Chen, Y. "Response to a Xinhua Correspondent's Questions on Problems of Market Prices of Commodities." Chinese Economic Studies, 15:3-4 (Spring/Summer 1982): 58-62.

403. Tian, J. "Implement the Reform of the Price System Vigorously and Reliably." Chinese Economic Studies, 18:4 (Summer 1985): 87-100.

404. "Prices Continue to Rise on Spot, Future Delivery." Daily News Record, 13 (Sept. 30, 1983); 2.

405. "Chinese Prices: Free Markets Exist--Official." The Economist, 284 (Sept. 25, 1982): 89.

406. Delfs, R. "High Cost of Stable Prices." The Far Eastern Economic Review, 115 (March 12-18, 1982): 84-86.

407. "Moving to the Market: China Rides a Tricky Bicycle."
     New York Times, 134 (Aug. 19, 1985): A19.

408. Bennett, A. "Peking Is Finding It Difficult to Let Prices
     Float." The Wall Street Journal (Feb. 22, 1985).

National Wealth and Balance Sheets
[224]

409. "Economy Moves Towards Balance." Beijing Review, 26
     (Oct. 17, 1983): 4-6.

410. Zhu, C. "The Balance and Readjustment." Chinese Economic Studies, 16:1 (Fall 1982): 3-18.

411. Xu, D. "Opening Speech at the Symposium on the Theory
     of Overall Balance of the National Economy." Chinese
     Economic Studies, 16:1 (Fall 1982): 19-29.

412. Sun, Y. "A Discussion of Some Prerequisites to the
     Overall Balance of the National Economy." Chinese
     Economic Studies, 16:1 (Fall 1982): 41-52.

413. Wang, D.A. "A Comprehensive Balance Should Mean Overall
     Balance in the Process of Social Reproduction."
     Chinese Economic Studies, 16:2 (Winter 1982/83):
     51-66.

414. Xue, M. "Readjust the National Economy and Strike an
     Overall Balance." Chinese Economic Studies, 16:2
     (Winter 1982/83): 67-84.

415. Liu, G. "Some Issues Concerning the National Economic
     Overall Balance." Chinese Economic Studies, 16:2
     (Winter 1982/83): 85-106.

416. Ge, Z. "On the Issue of Balancing the State Budget."
     Chinese Economic Studies, 16:2 (Winter 1982/83):
     107-23.

Social Indicators and
Social Accounts
[225]

417. "On the 12th Party Congress: Scientifically Understand
     and Handle Class Struggle in China." Beijing Review,
     25 (Dec. 6, 1982): 16-21.

418. "Economic and Social Achievements in 1982." Beijing Review, 26 (May 9, 1983): 6-7.

419. "Communique on the Statistics of 1985 Economic and Social Development." Beijing Review, 29 (March 24, 1986): 27-33.

420. "The Reward System in China." International Studies of Management and Organization, 12 (Summer 1982): 77-89.

421. "Entrepreneurs in China Are Quick to Seize Opportunity." The Wall Street Journal (March 11, 1985).

422. McFarlane, B. "Political Economy of Class Struggle and Economic Growth in China, 1950-1982." World Development, 11 (Aug. 1983): 659-72.

Productivity and Growth:
Theory and Data
[226]

423. "China Aims for Increasing Industrial Output: Yet, Infrastructural Problems Could Be a Hindrance to Plans." American Banker, 147 (Dec. 21, 1982): 7.

424. Dirksen, E. "Chinese Industrial Productivity in an International Context." World Development, 11 (April 1983): 381-87.

DOMESTIC MONETARY AND FISCAL THEORY AND INSTITUTIONS
[300]

DOMESTIC MONETARY AND FINANCIAL THEORY AND INSTITUTIONS
[310]

425. Stefani, G. "Finance Locale, Investimente e Servizi Urbani in Cina." [Local Finance, Investment and Urban Services in China; Eng. summary] Bancaria, 37:5 (May 1981): 454-69.

426. Boreham, G.F. "Financial Development in Southeast Asia." Canadian Banker ICB Review, 89 (Aug. 1982): 6-8+.

427. Yian, Y.; Jin, R.; and Yuan, Z. "Further Reform and Improve the Financial System." Chinese Economic Studies, 17:3 (Spring 1984): 53-60.

428. Wang, B. "On Several Problems Involving Financial Work." Chinese Economic Studies, 17:4 (Summer 1984): 81-98.

429. Goodstadt, L. "Peking Elevates the Bank of China's Status." Euromoney (Sept. 1979): 66-67+.

430. Gigot, P. "Banking on the Future: Look Behind the Scenes." The Far Eastern Economic Review, 108 (May 16, 1980): 70+.

431. Loong, P. "Second Look at the Role of Socialist Banking." The Far Eastern Economic Review, 110 (Dec. 19-25, 1980): 40-42.

432. Delfs, R. "New Kind of Planning [The World Bank Report on China]." The Far Eastern Economic Review, 113 (Aug. 14-20, 1981): 48-50.

Domestic Monetary and Financial
Theory and Policy
[311]

433. "China's Bank Studies U.S. to Help Boost Development." American Banker, 150 (March 7, 1985): 20.

434. Zhang, E. "On Banking Reform." Beijing Review, 24 (July 20, 1981): 24-27.

435. "Major Reform in Banking System." Beijing Review, 26 (Oct. 17, 1983): 6-7.

436. "Banking Reform Favors Centralization." Beijing Review, 27 (April 9, 1984): 16-18.

437. Liu, H. "Regulating Economy with Monetary Policy." Beijing Review, 28 (Dec. 23, 1985): 23-25.

438. Zhang Zeyu. "Balancing Foreign Currency in Enterprises." Beijing Review, 29 (May 19, 1986): 4-5.

439. Walter, C.E. "Dual Leadership and the 1956 Credit Reforms of the People's Bank of China." The China Quarterly, 102 (June 1985): 277-90.

440. Liu, M. "Higher Interest Rates to Offset Cash Shortage." The Far Eastern Economic Review, 104 (April 6, 1979): 80-81.

441. Bonavia, D. "Joker in the Money Pack." The Far Eastern Economic Review, 107 (Feb. 29, 1980): 32.

442. Rowley, A. "Compromising a Gold Cache." The Far Eastern Economic Review, 108 (April 25, 1980): 85.

443. Delfs, R. "One Currency, One System." The Far Eastern Economic Review, 125 (Sept. 13, 1984): 68-70.

444. Rowley, A. "Another Modernization." The Far Eastern Economic Review, 127 (Jan. 10, 1985): 50-53.

445. Lee, M. "Monetary Policy: Change from the Top." The Far Eastern Economic Review, 128 (May 9, 1985): 72.

446. Delfs, R. "Peking Cuts Its Losses." The Far Eastern Economic Review, 133 (July 17, 1986): 50-51.

447. DeWulf, L. "Financial Reform in China." Finance Development, 22:4 (Dec. 1985): 19-22.

448. Imai, H. "China's New Banking System: Changes in Monetary Management." Pacific Affairs, 58 (Fall 1985): 451-72.

449. "New Central Bank Chief Is Appointed in China." The Wall Street Journal (March 22, 1985).

Commercial Banking
[312]

450. Yan Kalin. "How to Deal with Losing Enterprises." Beijing Review, 28 (March 11, 1985): 25-26.

451. Sun Ping. "Bank Offices Active in Beijing." Beijing Review, 29 (June 2, 1986): 27-28.

452. "Let a Hundred Branches Bloom--Well, a Dozen Maybe." The Economist, 270 (March 10, 1979): 107-8.

453. "Foreign Banks in China: Toe in a Hold." The Economist, 295 (May 4, 1985): 83-84.

454. "The New, Aggressive Bank of China." Euromoney (July 1983): 128-29+.

455. "Hanging in There to Show a Commitment." The Far Eastern Economic Review, 111 (March 27-April 2, 1981): 66-68.

456. "Checking Accounts Make a Comeback in China." The Wall Street Journal (Oct. 30, 1984).

457. "China Lets 2 Provinces Open Their Own Banks." The Wall Street Journal (May 29, 1985).

Capital Markets
[313]

458. Chang, H. "The 1982-83 Overinvestment Crisis in China." Asian Survey, 24 (Dec. 1984): 1275-1301.

459. Loong, P. "Have Dollars, Will Sell." The Far Eastern Economic Review, 11 (Jan. 9-15, 1981): 53-54.

460. Delfs, R. "Lenders Dwarfing All Competitors." The Far Eastern Economic Review, 120 (May 5, 1983): 72.

461. Lee, M. "Green Light for a New Red Bourse." The Far Eastern Economic Review, 127 (March 28, 1985): 67-69.

462. Delfs, R. "That Unmentionable Word." [equity financing for large enterprises] Far Eastern Economic Review, 130 (Nov. 21, 1985): 136-37.

463. Berges, A.; McConnell, J.J.; and Schlarbaum, G.G. "The Turn-of-the-Year in China." Journal of Finance, 39:1 (March 1984): 185-92.

464. "China Alters Course with Sale of Stock." New York Times, 134 (Feb. 11, 1985): D4.

Credit to Business, Consumer, etc.
[314]

465. "Credit Card Fraud Sprouts in China." The Wall Street Journal (Jan. 8, 1985).

FISCAL THEORY AND POLICY; PUBLIC FINANCE
[320]

466. "Tax Evaders Cut State Revenue." Beijing Review, 28 (June 3, 1985): 9-10.

467. Wang Dacheng. "China Reigns in Disturbing Deficit." Beijing Review, 29 (Feb. 24, 1986): 4-5.

468. "Who Enjoys Subsidies? Who Pays?" Beijing Review, 29 (Aug. 11, 1986): 28-29.

469. Jao, J.C. "Recent Developments in China's Tax System." Bulletin for International Fiscal Documentation, 35:1 (Jan. 1981): 16-23.

470. "China: The New Tax Laws Offer Few Incentives." Business Week (Sept. 29, 1980): 52+.

Fiscal Theory and Policy; Public Finance 45

471. Xu, R. "A Preliminary Discussion of Comprehensive Public Finance." *Chinese Economic Studies*, 16:2 (Winter 1982/83): 20-28.

472. Dai, Y. "An Investigation of Fiscal Subsidy." *Chinese Economic Studies*, 18:4 (Summer 1985): 71-77.

473. "At War Over China: Asian Development Bank." *The Economist*, 286 (Feb. 26, 1983): 84.

474. Rowley, A. and Loong, P. "Testing of the Waters." [government bonds] *The Far Eastern Economic Review*, 107 (Jan. 11, 1980): 50-51.

475. Srodes, J. "Rong Seeks the Right Direction." *The Far Eastern Economic Review*, 109 (July 11, 1980): 48.

476. "Individual Income Tax Law." *The Far Eastern Economic Review*, 109 (Sept. 26, 1980): 52-53.

477. Loong, P. "China Leaves Some Grey Areas." [income tax] *The Far Eastern Economic Review*, 109 (Sept. 26, 1980): 108+.

478. "A Tax Upon Business." *The Far Eastern Economic Review*, 128 (May 30, 1985): 84.

479. Shih, A. and Au-Yeung, P.K. "Revenue Law and Practice in the People's Republic of China." *International Fiscal Documents*, 37:3 (March 1983): 99-104.

480. Kornai, J. "The Soft Budget Constraint." *Kyklos*, 39:1 (1986): 3-30.

481. Brauchli, M. "Chinese Exchange Controls, Latest Tax Tied to Worries About Inflation, Reserves." *The Wall Street Journal* (May 16, 1985).

482. "China's New Tax Law Apparently Misread." *The Wall Street Journal* (May 20, 1985).

INTERNATIONAL ECONOMICS
[400]

INTERNATIONAL TRADE THEORY
[410]

483. Zhao Ziyang. "Expanding Economic Exchanges and Promoting Common Prosperity." Beijing Review, 27:suppl. (June 18, 1984): v-vii.

484. Li Honglin. "Open Policy Essential to Socialism." Beijing Review, 28 (April 1, 1985): 15-18.

485. Chen Qiwei. "Why Is China Opening to the Outside?" Beijing Review, 28 (April 1, 1985): 18-22.

486. "Readjustment Policies Likely to Continue; Trade, Investment Prospects Remain Bright." Business America, 4 (Oct. 5, 1981): 17-20.

487. Zhao, Y. "On Estimation of the Gain and Loss of Foreign Trade." Challenge, 26:4 (Sept./Oct. 1983).

488. Ranis, G. "China's Open Door Is Open for Good." [editorial] Challenge, 28 (Nov./Dec. 1985): 59-60.

489. Ch'iang, L. "Distinguish the Correct Line from the Incorrect Ones; Actively Develop Socialist Foreign Trade." Chinese Economic Studies, 12:1-2 (Fall/Winter 1978/79): 83-97.

490. Cheng, Y. "Manage Foreign Trade Work Well." Chinese Economic Studies, 15:3-4 (Spring/Summer 1982): 139-43.

491. Zou, S. "The Problem of Export Strategy." Chinese Economic Studies, 16:3 (Spring 1983): 10-23.

492. Yuan, W. and Wang, J. "We Must Review and Reevaluate the Role of Foreign Trade in the Development of the National Economy." Chinese Economic Studies, 16:3 (Spring 1983): 24-39.

493. Wang, L. "On the Role of Foreign Trade Under Socialism." Chinese Economic Studies, 16:3 (Spring 1983): 48-65.

494. Xu, S. "On the Development of China's Foreign Trade." Chinese Economic Studies, 16:4 (Summer 1983): 3-16.

495. Zhang, L. "Reform Blazes a New Path for Foreign Trade." Chinese Economic Studies, 16:4 (Summer 1983): 17-26.

496. He, X. "Exploit the Role of Foreign Trade and Accelerate the Rate of China's Economic Development." Chinese Economic Studies, 16:4 (Summer 1983): 37-50.

497. Wang, K. "Tie Industry with Trade: An Important Way to Develop Foreign Trade." Chinese Economic Studies, 16:4 (Summer 1983): 70-77.

498. Li, H. "Socialism and Opening Up to the Outside World." Chinese Economic Studies, 19:1 (Fall 1985): 26-39.

499. Wang, J. "International Trade Engineering Is an Engineering Science." Chinese Economic Studies, 19:1 (Fall 1985): 81-88.

500. Bonavia, D. "Superpower Links Are the Prime Concern." The Far Eastern Economic Review, 127 (March 21, 1985): 92+.

501. Wassermann, U. "Zurich Conference on Trade with China." Journal of World Trade Law, 15:6 (Dec. 1981): 553-57.

502. Ishimine, T. "Organization, Ideology and Performance of China's International Trade." Malayan Economic Review, 23:2 (Oct. 1978): 1-15.

TRADE RELATIONS, COMMERCIAL POLICY, INTEGRATION
[420]

503. Reynolds, B.L. "Economic Reforms and External Imbalance in China, 1978-81." The American Economic Review, 73 (May 1983): 325-28.

504. Suharchuk, G.D. "Modernization in China and Foreign Policy." Asian Survey, 24 (Nov. 1984): 1157-62.

505. "Heilongjiang Improves Foreign Trade." Beijing Review, 29 (Aug. 4, 1986): 28-29.

506. Yang, D.J. and Shao, M. "China's Push for Exports Is Turning into a Long March." Business Week (Sept. 15, 1986): 66+.

507. Ranis, G. "China's Open Door Is Open for Good." Challenge, 28:5 (Nov./Dec. 1985): 59-60.

508. "China: Joining In." The Economist, 298 (Jan. 18, 1986): 58-59.

509. Delfs, R. "Economic Monitor, Export Picture Brightens." The Far Eastern Economic Review, 133 (July 31, 1986): 80-81.

510. Zhao Ziyang. "The Objectives of China's Foreign Policy: For Lasting Peace, Increased Friendly Cooperation and Co-Prosperity." International Affairs, 61 (Autumn 1985): 577-80.

511. Dirksen, E. "Chinese Industrial Productivity in an International Context." World Development, 11:4 (April 1983): 381-87.

Trade Relations
[421]

512. "Chinese International Trade." Accountant (London), 183 (Sept. 25, 1980): 503-4.

513. "Market Opens Wide to Foreign Technology." Advertising World, 11 (Sept. 1984): 30+.

514. Surls, F.M. "New Directions in China's Agricultural Imports." American Journal of Agricultural Economics, 62:2 (May 1980): 349-55.

515. Kilpatrick, J.A. "Chinese Agriculture: Development, Production, and Trade." [discussion] American Journal of Agricultural Economics, 62:2 (May 1980): 359-61.

516. Kraar, L. "Beijing's Overseas Chinese Connection." Asia, 2 (July 1979): 4-7.

517. Kueh, Y.Y. "China's Food Balance and the World Grain Trade; Projections for 1985, 1990 and 2000." Asian Survey, 24 (Dec. 1984): 1247-74.

518. Yuanzheng, L. "The Management of China's Modernisation and Its Impact on the Rest of the World." Australian Journal of Management, 7:1 (June 1982): 1-8.

519. Yan, X. "World Economy Ahead." Beijing Review, 24 (March 23, 1981): 15-19+.

520. "Economic Co-Operation with Foreign Countries." Beijing Review, 24 (June 1, 1981): 9-10.

521. "On China's Economic Relations with Foreign Countries." Beijing Review, 25 (May 31, 1982): 13-16.

522. Chen Muhua. "Prospects for China's Foreign Trade in 1983." Beijing Review, 26 (Feb. 7, 1983): 14-17.

523. "Opening to the Outside World and Self-Reliance." Beijing Review, 26 (March 14, 1983): 14-20.

524. "Trade Balance Tips into the Red." Beijing Review, 28 (Feb. 4, 1985): 6-7.

525. "Trade Deficit on the Rise." Beijing Review, 28 (Aug. 5, 1985): 28-29.

526. Gullo, D.T. "Prospects for China Trade Through 1985." Business America, 3 (Aug. 11, 1980): 3-16.

527. "China Exhibition Expected to Increase Two-Way Trade: $10 Billion by 1985." Business America, 3 (Oct. 6, 1980): 14-15.

528. Verzariu, P. "Update on Countertrade with China." Business America, 5 (Jan. 11, 1982): 8-10.

529. Chen, N.-R. and Lee, J. "New Economic Reforms Create More Opportunities." Business America, 8 (March 4, 1985): 28-29.

530. Chen, N.-R, and Monk, L.B. "Five Year Plan Offers New Business Prospects." Business America, 9 (March 17, 1986): 24-25.

531. Marer, P. "Future for Trade with China." Business Horizons, 22 (April 1979): 6-13.

532. Ichimura, S. "The Time to Reassess China's Influence Is Now." [editorial] Business Japan, 29 (Dec. 1984): 7.

533. Young, L.H. "China's Long Road to Large-Scale Trade." Business Week (May 28, 1979): 132-33.

534. "Hidden Hazards in a China Grain Pact." Business Week (Oct. 27, 1980): 58-59.

535. Deyan, Z. "China Emerging as a Trading Nation." The Canadian Business Review, 12 (Spring 1985): 56-62.

536. Mozingo, D. "West's Stake in China's Economic Struggle." Center Magazine, 15 (May/June 1982): 9-16.

537. "Favored Nation Designation for Red China under Study." Chemical Marketing Reporter, 215 (April 30, 1979): 14+.

538. Chao, K. "China Watchers Tested." The China Quarterly, 81 (March 1980): 97-104.

539. Quihua, Q. "Research on the World Economy in China." The China Quarterly, 84 (Dec. 1980): 720-26.

540. Stuart, D.T. and Tow, W.T. "Chinese Military Modernization: The Western Arms Connection." The China Quarterly, 90 (June 1982): 253-70.

541. Wang, G.C. "Issues in China's International Trade." Chinese Economic Studies, 16:3 (Spring 1983): 6-9.

542. "World Scans China Turnaround: New Economic Tilt Could Affect Many Relationships." Christian Science Monitor, 76 (Nov. 5, 1984): 1.

543. "West Meets East at China Trade Fair." Christian Science Monitor, 77 (May 7, 1985): 11.

544. "China Trade: A Blitz in Textiles." Citibank (June 1981): 4-6.

545. Tsumuri, Y. "Your Check List for an Approach to China." The Columbia Journal of World Business, 14 (Summer 1979): 8-15.

546. "Some Answers on the China Trade." The Columbia Journal of World Business, 14 (Summer 1979): 51-64.

547. Tung, R.L. "Corporate Executives and Their Families in China: The Need for Cross-Cultural Understanding in Business." The Columbia Journal of World Business, 21 (Spring 1986): 21-25.

548. Crane, A.T. and Suttmeier, R.P. "Nuclear Trade with China." The Columbia Journal of World Business, 21 (Spring 1986): 35-40.

549. Wiesemeyer, J. "Finding Balance in Trading with the People's Republic of China." Commodities, 10 (May 1981): 35-37.

550. "Exports: Promises, Promises." The Economist, 270 (March 10, 1979): 87-88.

551. "Some Reds Are More Equal Than Others." The Economist, 276 (Sept. 20, 1980): 93-94.

552. "China: Comrades Go Shopping." The Economist, 291 (June 9, 1984): 34-35.

553. "New Economic Ties." The Economist, 293 (Nov. 10, 1984): 79.

554. "Trade Still Takes a Slow Boat to China." The Economist, 294 (Feb. 16, 1985): 65-66.

555. "Trade Flows Where No Diplomat Goes." The Economist, 298 (March 22, 1986): 74-75.

556. Snyderman, N. "Industry Observer." Electronic News, 29:suppl. (Nov. 21, 1983).

557. "China Trade Openings for Engineers." Engineering News-Record, 212 (May 10, 1984): 10-11.

558. "China's Trade Door Squeaks Open." Engineering News-Record, 213 (Nov. 22, 1984): 24-25+.

559. Goodstadt, L. "Is China Awakening at Last?" Euromoney (July 1984): 132+.

560. Liu, M. "U.S. and Peking Find a Way: Putting Things in Proportion." The Far Eastern Economic Review, 104 (May 25, 1979): 73-74.

561. Ram, M. "Rivals in the Market." The Far Eastern Economic Review, 113 (July 24-30, 1981): 41-42.

562. Ma, T. "Chinese Screen." The Far Eastern Economic Review, 117 (Oct. 10-16, 1982): 73.

563. Bonavia, D. "In the Driving Seat." The Far Eastern Economic Review, 124 (May 10, 1984): 16-17.

564. Chanda, N. and Manning, R. "Congress Goes Critical." The Far Eastern Economic Review, 124 (June 28, 1984): 66-68.

565. do Rosario, L. "Trading Patterns See Some Dramatic Shifts." The Far Eastern Economic Review, 127 (March 21, 1985): 78-79.

566. do Rosario, L. "Trading on a Reputation (Sinochem)." The Far Eastern Economic Review, 131 (Jan. 2, 1986): 50-51.

567. do Rosario, L. "Long March to Sufficiency." [steel imports] The Far Eastern Economic Review, 132 (May 1, 1986): 54-55.

568. Hickok, S.A. and Arguelles, R. "China's Rapid Trade Growth and Impact on the World Economy." Federal Reserve Bank of New York Quarterly Review, 7:4 (Winter 1982/83): 41-51.

569. Kolbenschlag, M. "China Trade." Forbes, 125 (June 9, 1980): 37-38.

570. Kraar, L. "China's Narrow Door to the West." Fortune, 99 (March 26, 1979): 62-69.

571. Anand, V. "Intrepid Entrepreneurs Discover Markets in China." Global Trade Executive, 104 (Feb. 1986): 20-21.

572. Pye, L.W. "The China Trade: Making the Deal." Harvard Business Review, 64 (July/Aug. 1986): 74+.

573. Hendryx, S.R. "The China Trade: Making the Deal Work." Harvard Business Review, 64 (July/Aug. 1986): 75+.

574. Donath, B. "China Trade Growth: Will the Cookie Grumble?" Industrial Marketing, 65 (Nov. 1980): 52-54.

575. Samli, C. and Kosenko, R. "Support Service Is Key for Technology Transfer to China." Industrial Marketing Management, 11 (April 1982): 95-103.

576. Dong Fureng. "Some Problems Concerning China's Strategy in Foreign Economic Relations." International Social Science Journal, 25:3 (1983): 455-67.

577. Phanachet, U. and Huixiang, Z. "Techniques for Exporting to China." International Trade Forum, 19 (March 1983): 8-11+.

578. "China Slaps Higher Prices on Textiles." Journal of Commerce and Commercial, 360 (April 25, 1984): 1A.

579. "Trade Gap Grows Wider in China." Journal of Commerce and Commercial, 365 (July 26, 1985): 1A.

580. "The International Treatment of State Trading." Journal of World Trade Law, 16 (Nov./Dec. 1982): 480-96.

581. Wassermann, U. "China's Expanding Trade." Journal of World Trade Law, 19:5 (Sept./Oct. 1985): 542-46.

582. Warrington, M.B. and McCall, J.B. "Negotiating a Foot into the Chinese Door." Management Decision, 21:2 (1983): 3-13.

583. Szuprowicz, B.O. "China Fever: Scrambling for Shares in a $600 Million Buying Spree." Management Review, 68 (May 1979): 8-16.

584. Denis, R. and Munson, S. "Trading with China: A Boon for Some, a Disappointment for Many." Management Review, 72 (May 1983): 13-20.

585. Aaron, B. "China: A Seller's Market." Nation's Business, 68 (April 1980): 24-26+.

586. "China's 'Open Door' to West Begins to Close." New York Times, 134 (Aug. 4, 1985): 1.

587. "An Emerging China's Impact." New York Times, 134 (Aug. 21, 1985): D2.

588. "China's Place in World Trade." The OECD Observer, 114 (Jan. 1982): 10-11.

589. "China to Import More Technology, Know-How." [M.W. Kellogg Co.] Oil & Gas Journal, 82 (Nov. 26, 1984): 58-59.

590. Solinger, D.J. "Commercial Reform and State Control: Structural Changes in Chinese Trade, 1981-1983." Pacific Affairs, 58 (Summer 1985): 197-215.

591. Ishimine, T. "Organization, Ideology, and Performance of China's International Trade." Pakistan Economic and Social Review, 17:1-2 (Spring/Summer 1979): 1-22.

592. "Chinese Promote Frozen Food at 15th Munich Food Fair." Quick Frozen Foods International, 26 (Jan. 1985): 129.

593. "China's Planned Acquisition of Sophisticated Western Technology Stirs Debate over Export Controls." Research & Development, 27 (Feb. 1985): 122+.

594. "China's Free-Market Tilt: Good News and Bad News for Foreign Companies." The Wall Street Journal (Nov. 8, 1984).

595. "GATT Grants China's Bid for Observer Status." Women's Wear Daily, 148 (Nov. 7, 1984): 24.

Commercial Policy
[422]

596. Knowles, H.A. "Chinese Reassess Trademark Attitudes." Advertising Age, 50 (May 14, 1979): sec. 2:S3.

597. Callahan, J.M. "China! A Trade Dynasty Is Born."
     Automotive Industry, 165 (Nov. 1985): 58-60.

598. Mann, P. "China Export Policy Takes Final Form."
     Aviation Week & Space Technology, 116 (Jan. 25, 1982):
     57-58.

599. "Reforming the Foreign Trade Structure." Beijing Review,
     27 (Oct. 22, 1984): 4-5.

600. "Development and Reform in Commerce." Beijing Review,
     28 (July 29, 1985): 25-27.

601. Wang, D. "Open Policy to Remain in Force." Beijing
     Review, 28 (Aug. 26, 1985): 4-5.

602. Liu, H. "Technology Import Reaches New High." Beijing
     Review, 29 (March 10, 1986): 22-24.

603. Yuan Zhenmin. "China Adopts Law on Foreign Enterprises."
     Beijing Review, 29 (May 5, 1986): 14-17.

604. "China's New Trademark Policy Is Another Sign of Its
     Interest to Enter World Trading System." Business
     America, 2 (May 7, 1979): 9.

605. Mundheim, R. "Claims/Assets Agreement with the People's
     Republic of China: What It Means, How It Was Nego-
     tiated." Business America, 2 (July 16, 1979): 7-9.

606. Chen, N.R. and Lee, J.L. "US-PRC Agreement, MFN Support
     Continued Growth of Trade." Business America, 3
     (July 28, 1980): 17-18.

607. Seidman, H.L. "Commerce's Export Programs for the
     1980s." Business America, 3 (Oct. 20, 1980): 22-24.

608. Lee, J.L. "China Continues to Emphasize Trade and
     Foreign Investment." Business America, 5 (Feb. 8,
     1982): 44-45.

609. Chen, N.-R. and Lucyk, C.L. "China's Premier Zhao
     Reaffirms Commitment to Open Door Policy." Business
     America, 7 (Feb. 20, 1984): 39-40.

610. Jones, D. "Power Play Behind China's Trade Fiasco."
     Business Week (March 9, 1981): 38-39.

611. "The Long Road to China Is Paved in Red Tape." Chemical Week, 131 (Jan. 19, 1983): 57-58.

612. "Compensation Trade: The China Perspective." China Business Review, 9 (Jan./Feb. 1982): 50-52.

613. "The Textiles Deadlock." China Business Review, 9 (Nov./Dec. 1982): 31-35.

614. Kueh, Y.-Y. and Howe, C. "China's International Trade; [Policy and organizational change, their place in the economic readjustment]." The China Quarterly, 100 (Dec. 1984): 813-48.

615. Zhao, Y. "On Export-Import Procedure." Chinese Economic Studies, 16:3 (Spring 1983): 66-75.

616. Cai, T. "Conventional Practice in Setting Export-Import Prices." Chinese Economic Studies, 16:3 (Spring 1983): 85-92.

617. Chongwei, J. "China's Utilization of Foreign Funds and Relevant Policies." Chinese Economic Studies, 17:2 (Winter 1983/84): 37-49.

618. "EEC, China Hit U.S. Export Subsidies Complaint." Daily News Record, 13 (Oct. 12, 1983): 3.

619. "Foreign Economic Contract Law: A Breakthrough." East Asian Executive Reports, 7 (April 15, 1985): 9.

620. Erlich, P. "China CPU Needs Outlined." Electronic News, 30:suppl. (Nov. 5, 1984).

621. Rowley, A. "Cracking the Cryptic Code: China's Law on Joint Ventures." The Far Eastern Economic Review, 105 (July 20, 1979): 49-52.

622. Wilson, P. "Change in Trade Policy." The Far Eastern Economic Review, 108 (June 20, 1980): 45-46.

623. Delfs, R. "Flirting with Freer Trade [GATT]." The Far Eastern Economic Review, 117 (Sept. 24-30, 1982): 126.

624. Nations, R. "Raising the Barriers." [Liberalize restrictions on exports of sophisticated technology] The Far Eastern Economic Review, 120 (June 16, 1983): 16+.

625. "The Politics of Trade." The Far Eastern Economic Review, 129 (Sept. 5, 1985): 16-18.

626. "China Blasts New U.S. Rule on Textiles." Journal of Commerce, 363 (March 8, 1985): 1A.

627. Chu, L. "Doing Business with China: How Peking's New Joint Ventures Law Will Work." Management Review, 69 (July 1980): 59-61.

628. Lowe, J.Y. "China Modernizing Business Laws to Boost Trade and Investment." Marketing Review, 69 (April 1980): 31-32.

629. Crow, P. "Chinese Hot Potatoe." [Tariffs on imports of unfinished gasoline] Oil & Gas Journal, 82 (July 2, 1984): 39.

630. Meilach, Dona Z. "Peking Reverses Open-Door Policy on Computer Imports." PC Week, 2 (Sept. 17, 1985): 137.

631. Alford, E.P. "Law and Chinese Foreign Trade." [review article] Problems of Communism, 28 (Sept./Dec. 1979): 81-84.

632. "Westerners' Joint Ventures in China Encounter Bureaucratic Delays, Foreign Exchange Blocks." The Wall Street Journal (March 18, 1985).

633. Bennett, Amanda. "China Tightens Grip on Its Economy, But Foreign Trade Appears Unaffected." The Wall Street Journal (March 20, 1985).

634. Namiotkiewicz, W. "Foreign-Policy Aspects of the Four Modernizations." World Marxist Review, 23 (April 1980): 65-68.

International Economic Integration
[423]

635. Zhang, Z. "China Positions Itself to Rejoin GATT." Beijing Review, 29 (March 10, 1986): 4-5.

636. "China Seeks Seat on ADB Board." Beijing Review, 29 (May 26, 1986): 8.

637. Chen, N.-R. "China's Foreign Trade/Economy Slowed in First Half of Year." Business America, 9 (Aug. 18, 1986): 24-25.

638. Klatt, W. "Global China." [review article] The China Quarterly, 89 (March 1982): 105-9.

639. Lee, R.W. "Political Absorption of Western Technology: The Soviet and Chinese Cases." Studies in Comparative Communism, 15 (Spring/Summer 1982): 9-33.

Trade Relations with the U.S.
[424]

640. Asher, J. "China: The Trade Wall Is Coming Down." ABA Banking Journal, 71 (July 1979): 28-30+.

641. "New American Dream--It's Doing Business with China." Air Conditioning, Heating and Refrigerating News, 146 (Jan. 29, 1979): 60-62.

642. King, R., Jr. "Facing Hard Facts in U.S.-PRC Container Trade." American Shipper, 25 (April 1983): 11-14.

643. King, R., Jr. "Behind the U.S.-China Impasse." American Shipper, 26 (Feb. 1984): 16-27.

644. Chyba, C.F. "U.S. Military-Support Equipment Sales to the People's Republic of China." Asian Survey, 21 (April 1981): 469-84.

645. Lukin, V. "Relations Between the U.S. and China in the 1980s." Asian Survey, 24 (Nov. 1984): 1151-56.

646. Simon, D.F. "The Challenge of Modernizing Industrial Technology in China: Implications for Sino-U.S. Relationships." Asian Survey, 26 (April 1986): 420-39.

647. Mann, P. "China, U.S. Set Further Arms Buy Talks." Aviation Week & Space Technology, 120 (June 11, 1984): 22-23.

648. Brody, M. "Naked Protectionism; Quotas on Clothing Imports Chill Producers, Consumers Alike." [editorial] Barron's, 63 (July 25, 1983): 11.

649. Scheibla, S.H. "China Trade: It Holds Dangers, an Expert Warns, for American Firms." [Interview with Ray Cline] Barron's, 63 (Oct. 24, 1983): 52+.

650. "U.S. Court Trial Violates International Law." Beijing Review, 26 (March 14, 1983): 24-27+.

651. "More Agreements Reached with U.S." Beijing Review, 27 (May 21, 1984): 11-12.

652. "China Protests U.S. Textile Rule." Beijing Review, 27 (Aug. 27, 1984): 7-8.

653. "A Move Towards Protectionism." Beijing Review, 28 (March 25, 1985): 15.

654. "Chemical Bank Expands Business." Beijing Review, 28 (June 3, 1985): 34-35.

655. Xiao, X. "Sino-US Trade Ties and Legal Exchanges." Beijing Review, 28 (July 8, 1985): 20-22.

656. "Sino-U.S. Giant Coal Mine Pact Inked." Beijing Review, 28 (July 8, 1985): 29-30.

657. Zi, Z. and Zhuang, Q. "Sino-U.S. Relations: Opportunities and Potential Crisis." Beijing Review, 28 (Oct. 14, 1985): 21-26.

658. "Entering the China Market." Business America, 2 (Feb. 26, 1979): 2-7.

659. "Secretary Kreps Reports on China Trip." Business America, 2 (June 4, 1979): 6-8.

660. Chen, N.R. "Spotlight on US-PRC Trade: Review of 1978 Transactions, and Prospects for Expansion." Business America, 2 (Sept. 24, 1979): 8-10.

661. "U.S.-China Exchanging Trade Shows." Business America, 2 (Nov. 5, 1979): 2-7.

662. Maffry, A., Jr. "U.S. Exhibition Opens in Beijing." Business America, 3 (Nov. 17, 1980): 3-8.

663. "Doing Business with China: An Analysis of Economic Trends and Implications for U.S. Business." Business America, 3 (Nov. 17, 1980): 9-12.

664. "Beijing Exhibition Lays Groundwork for Expansion of US-China Trade." Business America, 3 (Dec. 15, 1980): 8-11.

665. Lee, J.L. and Chen, N.R. "US-PRC Trade Is Rising Rapidly Despite Economic Readjustment." Business America, 4 (Feb. 9, 1981): 21-22.

666. Denny, D.L. "Prospects for US-China Trade." Business America, 4 (June 15, 1981): 3-7.

667. DiFederico, E.M. "U.S.-China Trade: A Decade of Development and Prospects for Growth." Business America, 5 (June 28, 1982): 2-10.

668. Scouton, W. "Mission Strengthens Ties with China and Japan." Business America, 6 (June 13, 1983): 3-8.

669. "The Chinese Economy." [economic policy and U.S. trade] Business America, 6 (June 13, 1983): 9-11.

670. "U.S. Liberalizes Controls on Exports of Technology to China." Business America, 6 (Nov. 28, 1983): 33-34.

671. "Accord on Industrial and Technological Cooperation Between the U.S. and the People's Republic of China." Business America, 7 (Jan. 23, 1984): 24-25.

672. Woodward, P. "President's Trip Strengthens U.S.-China Commercial Ties." Business America, 7 (May 14, 1984): 17-19.

673. Chen, N.-R. and Lee, J.L. "Reagan Visit to China, Accords Support Expansion of U.S. Exports." Business America, 7 (Aug. 20, 1984): 39-40.

674. Wethington, O. "Presidential Trade Mission Reflects New Era in U.S.-China Trade." Business America, 7 (Oct. 15, 1984): 3-7.

675. "Secretary Baldridge Visits China, India and USSR." Business America, 8 (June 10, 1985): 2-15.

676. Woodward, P.L. "U.S. Firms Vie for China's Telecommunications Market." Business America, 8 (Aug. 19, 1985): 2-4.

677. Monk, L.B. "Fourth US-China Trade Meeting Results in Four New Agreements." Business America, 9 (July 7, 1986): 19.

678. Hester, S.B. and Hinkle, D.E. "Balancing Its Trade Interests: The Impact of Chinese Textile Imports on the Market Shares of American Manufacturers." Business Economics, 19:3 (April 1984): 27-34.

679. Maher, P. "Outlook Brightens for Trade with China." Business Marketing, 69 (July 1984): 104+.

680. "Why Business with Beijing Will Get Better." Business Week (Sept. 26, 1983): 41-42.

681. "Keeping the Momentum Going After Reagan's Visit." Business Week (May 14, 1984): 55+.

682. Richman, B. "Sino-American Economic Relations: Constraints, Opportunities, and Prospects." California Management Review, 21 (Winter 1978): 13-28.

683. Webber, D. "How to Make Chemical Deals with the Chinese." Chemical Business (March 10, 1980): 9-12+.

684. "Negotiators with Chinese Need an Awareness of Pitfalls." Chemical Marketing Reporter, 215 (May 28, 1979): 14+.

685. "More Chinese Will Train in U.S." Chemical Week, 125 (Sept. 15, 1979): 70.

686. "Textile-Import Accord--Maybe." Chemical Week, 133 (Aug. 10, 1983): 9-10.

687. Trewhitt, J. and Dunphy, J.F. "China's Welcome for Chemicals." Chemical Week, 137 (Nov. 27, 1985): 13-14.

688. Schroeder, P.E. "Ohio-Hubei Agreement: Clues to Chinese Negotiating Practices." The China Quarterly, 91 (Sept. 1982): 486-91.

689. Tsurumi, Y. "US-China Trade: Prologue." The Columbia Journal of World Business, 14 (Summer 1979): 5-7.

690. Ruggles, R.L., Jr. "The Environment for American Business Ventures in the People's Republic of China." The Columbia Journal of World Business, 18 (Winter 1983): 67-73.

691. Shi, J. "Future Prospects for Broadening US-China Economic and Trade Cooperation." The Columbia Journal of World Business, 20 (1985): 57-58.

692. Wang, N.T. "United States and China: Business Beyond Trade--An Overview." The Columbia Journal of World Business, 21 (Spring 1986): 3-4.

693. Ross, M.C. "China and the United States' Export Controls System." The Columbia Journal of World Business, 21 (Spring 1986): 27-33.

694. Stoltenberg, C.D. "Trends in People's Republic of China Import Cases under US Trade Law." The Columbia Journal of World Business, 21 (Spring 1986): 41-48.

695. Mun, K.C. and Chan, T.S. "The Role of Hong Kong in United States-China Trade." The Columbia Journal of World Business, 21 (Spring 1986): 67-73.

696. "U.S., China Reach Pact on 9 Disputed Quotas." Daily News Record, 15 (March 12, 1985): 2.

697. "Reagan Orders Sweeping Curbs on Textile Imports from China." Discount Store News, 23 (Jan. 9, 1984): 2.

698. Adkins, L. "Capitalism Comes to China." Dun's Review, 115 (Jan. 1980): 74-75+.

699. "Two Most Favoured Nations? China and Russia." The Economist, 270 (March 10, 1979): 47-48.

700. "America Needles Hong Kong." [with Chinese textile exports] The Economist, 289 (Dec. 17, 1983): 66+.

701. Barty, E. "U.S. Computers: On a Slow Boat to China via Hong Kong." Electronic Business, 10 (March 1984): 176+.

702. "Carter Policy Makers Express High Hopes for Trade with China." Electronic News, 25:suppl. (Dec. 3, 1979).

703. Schwartz, L. "Plan to Ease Rules on Exports to China." Electronic News, 29:suppl. (Oct. 24, 1983).

704. Srodes, J. "Counting on Each Other." The Far Eastern Economic Review, 102 (Dec. 29, 1978): 41-42.

705. Williams, T. "Those Frozen Assets Were the Crux of It All." The Far Eastern Economic Review, 103 (March 16, 1979): 64+.

706. Srodes, J. "Easing the Way for China Deals." The Far Eastern Economic Review, 106 (Nov. 23, 1979): 66.

707. "Peking Puzzle for the Americans." The Far Eastern Economic Review, 107 (Feb. 15, 1980): 56-57.

708. Cohen, J.A. "Building Up the Joint Economic Framework." The Far Eastern Economic Review, 107 (March 7, 1980): 41+.

709. Smil, V. "US Should Be China's Key Technological Provider." The Far Eastern Economic Review, 107 (March 7, 1980): 66-68.

710. Westlake, M. "Scrambling for the Market." The Far Eastern Economic Review, 108 (May 2, 1980): 55-56.

711. Delfs, R. and Ma, T. "Tiger's Roar Is Real." The Far Eastern Economic Review, 119 (Feb. 3, 1983): 39-41.

712. Manning, R. "Chinese Puzzle." The Far Eastern Economic Review, 122 (Nov. 24, 1983): 74+.

713. Langston, N. "Open Pit, Closed Deal." [U.S. investment] The Far Eastern Economic Review, 129 (July 11, 1985): 66+.

714. Chanda, N. "New Planes from Old: U.S. Near Signing Contract for Advanced Avionics for China." The Far Eastern Economic Review, 131 (Jan. 2, 1986): 11-12.

715. do Rosario, L. "Jeep on a Bumpy Road." The Far Eastern Economic Review, 132 (June 12, 1986): 132-34.

716. Pearlstine, N. and Ling, F.S.H. "China Trade: A Note of Caution." Forbes, 123 (Feb. 5, 1979): 33-35.

717. Okita, S. "Japan, China and the United States: Economic Relations and Prospects." Foreign Affairs, 57:5 (Summer 1979): 1090-1110.

718. Oksenberg, M. "China Policy for the 1980s." Foreign Affairs, 59 (Winter 1980/81): 304-22; (Spring 1981): 939-40.

719. Kraar, L. "P.T. Barnum of China Trade (C. Abrams)." Fortune, 101 (May 5, 1980): 186-87+.

720. Bernstein, P.W. "Why Business Won't Follow Reagan to China." Fortune, 109 (May 14, 1984): 144.

721. Norton, R.E. "China Threatens to Shut the Trade Door." Fortune, 112 (Sept. 30, 1985): 119.

722. Ewing, H.G. "Eagle and the Dragon." Industrial Research and Development, 21 (May 1979): 138-42.

723. Miller, W.H. "China Trade: Euphoria Now and Big Dollars Later." Industry Week, 200 (Jan. 8, 1979): 17-18.

724. Miller, W.H. "New Foundation for US-China Trade." Industry Week, 201 (May 28, 1979): 19-21.

725. McClenahen, S. and Miller, W.H. "Tackling the China Challenge." Industry Week, 201 (May 28, 1979): 32-33+.

726. Miller, W.H. "US-China Trade: A Ten-Year Review." Industry Week, 212 (March 22, 1982): 93-94.

727. Miller, W.H. "U.S.-China Trade: Keeping the Momentum Building." Industry Week, 221 (May 28, 1984): 19-20.

728. Gray, P. and Dean, B.V. "Chinese-U.S. Symposium on Systems Analysis." Interfaces, 12 (Feb. 1982): 44-49.

729. "U.S.-China Representatives Initial Income Tax Accord." Journal of Accountancy, 157 (May 1984): 24-25.

730. "Export Clouds Lifting on U.S. Grain Horizon." Journal of Commerce, 357 (Aug. 3, 1983): 1A.

731. "US Mulls Closer Ties with China: Commission Explores Increased Sales of US Farm Products." Journal of Commerce and Commercial, 359 (Jan. 13, 1984): 1A.

732. "U.S. High-Tech Sales to China Likely to Soar." Journal of Commerce and Commercial, 360 (May 22, 1984): 1A.

733. "U.S. Groups Plan Projects for China." Journal of Commerce and Commercial, 362 (Dec. 4, 1984): 1A.

734. Petr, J.L. "Comparative Analysis of Thresholds of Non-Revolutionary Institutional Change: China and the

United States in the 1980s." Journal of Economic Issues, 20 (June 1986): 561-69.

735. Han, X. "China's Economic Reform and Sino-U.S. Relations." Journal of International Affairs, 39 (Winter 1986): 27-31.

736. "Interview: Paul D. Wolfowitz; U.S.-China Relations." Journal of International Affairs, 39 (Winter 1986): 33-39.

737. Tung, R.L. "U.S.-China Trade Negotiations: Practices, Procedures and Outcomes." Journal of International Business Studies, 13:2 (Fall 1982): 25-37.

738. Bennett, W.R. and Allen, B. "Opportunity for Small Business in the People's Republic of China." [editorial] Journal of Small Business Management, 17 (April 1979): 1-4.

739. Das, D.K. "The New U.S. Customs Regulation and China." Journal of World Trade Law, 19:3 (May/June 1985): 287-89.

740. Seixas, S. "Baubles and Bargains from the New China." Money, 8 (May 1979): 94-96.

741. Altmann, H. "A Frontier for U.S. Business." Nation's Business, 73 (Dec. 1985): 65-66.

742. Robinson, T.W. "Choice and Consequence in Sino-American Relations." Orbis, 25 (Spring 1981): 29-51.

743. Lipschutz, N. "China's Ambitious Modernization Plan Holds Promises and Questions for U.S." Paper Trade Journal, 163 (Oct. 15, 1979): 34-36.

744. Forger, G. "Growing Market for Plastics: Update on China." Plastics World, 37 (Nov. 1979): 54-56.

745. Martellaro, J.A. "China's Two Economies and the USA Factor." Rivista Internazionale di Scienze Economiche e Commerciali, 31:9 (Sept. 1984): 875-97.

746. Nevis, E.C. "Cultural Assumptions and Productivity: The United States and China." Sloan Management Review, 24:3 (Spring 1983): 17-29.

747. "U.S.-China Exchange Program." Telecommunications, 17 (Sept. 1983): 28+.

748. Sucheki, S.M. "Agreement with China Raises Imports." [special report] Textile Industries, 147 (Sept. 1983): 12.

749. Sucheki, S.M. "U.S.-China Agreement Deemed Unsatisfactory." Textile Industries, 147 (Oct. 1983): 38-39.

750. "U.S.-China Pact Raises New Doubts." Textile World, 133 (Sept. 1983): 23+.

751. Spielmann, P. "Markets: Tapping the Chinese." Venture, 7 (May 1985): 33-34.

752. Bennett, Amanda. "U.S. Firms Rush Through China's Open Door." The Wall Street Journal (April 8, 1985).

753. Merry, R.W. "Chinese Officials Turn the Tables on Visiting Vice President Bush." The Wall Street Journal (Oct. 17, 1985).

754. Sterba, James P. "The Sino-U.S. Relationship: Mistakes on Both Sides." The Wall Street Journal (Oct. 24, 1985).

755. "Two U.S. Firms Join China in Developing Peking Site." The Wall Street Journal (Sept. 12, 1985).

756. Wyllie, R.J.M., et al. "Bethlehem Cooperates in Big Iron Ore Project." World Mining, 32 (Oct. 1979): 107-11.

757. Ross, R.S. "International Bargaining and Domestic Politics: U.S.-China Relations Since 1972." World Politics, 38 (Jan. 1986): 256-87.

Trade Relations with
Other Countries
[425]

758. "Gaps in the Wall." [China and Great Britain] The Accountant, 183 (Oct. 9, 1980): 587-88.

759. "China in the Third World." Africa, 127 (March 1982): 69-70.

760. "Germans Win China Mill Contract." American Metal Market, 92 (Oct. 19, 1984): 3.

761. "Europe: Market Firm But Pricing Effect of Chinese Exports Is Uncertain." American Metal Market, 93 (Feb. 6, 1985): 8A.

762. "Krupp to Help Build Refinery for Chinese." American Metal Market, 93 (March 5, 1985): 2.

763. Bray, D. "Hong Kong: Its Economic Structure and Relationship with China." Asian Affairs, 11 (Oct. 1980): 293-300.

764. Kim, Shee Poon. "Politics of Thailand's Trade Relations with the People's Republic of China." Asian Survey, 21 (March 1981): 310-24.

765. Vertzberger, Y. "Political Economy of Sino-Pakistani Relations: Trade and Aid, 1963-82." Asian Survey, 23 (May 1983): 637-52.

766. Brick, P. "The Politics of Bonn-Beijing Normalization 1972-84." Asian Survey, 25 (July 1985): 773-91.

767. "China, West Germany Sign Trade Pact." Aviation Week, 111 (Oct. 29, 1979): 21-22.

768. "Developing Trade with Other Third World Countries." Beijing Review, 26 (Sept. 19, 1983): 15-18.

769. "Sino-German Nuclear Accord Signed." Beijing Review, 27 (May 21, 1984): 12.

770. "Hu and Deng Meet with Willy Brandt." Beijing Review, 27 (June 11, 1984): 8-9.

771. "FDR-China Ties Stressed by Zhao." Beijing Review, 27 (Oct. 15, 1984): 7-8.

772. "Foreigner Tabbed to Direct Factory." Beijing Review, 27 (Nov. 19, 1984): 8-9.

773. "Sino-Czechoslovak Trade Booms." Beijing Review, 27 (Dec. 10, 1984): 7.

774. "Arkhipov's Visit Boosts Trade, Improves Atmosphere." [USSR] Beijing Review, 28 (Jan. 7, 1985): 6-7.

Trade Relations, Commercial Policy, Integration 69

775. "Learn Update Management Skills." [West German technical assistance] Beijing Review, 28 (Jan. 21, 1985): 4.

776. "Sino-Soviet Border Trade Rejuvenated." Beijing Review, 28 (Jan. 28, 1985): 30-31.

777. "Portugal Interested in China Market." Beijing Review, 28 (May 27, 1985): 31-32.

778. "From Ice Cream to Nuclear Power." [Western Europe] Beijing Review, 28 (June 3, 1985): 16-17.

779. Wei, Y. "Boosting Sino-Latin American Trade Relations." Beijing Review, 28 (Oct. 28, 1985): 15-17.

780. "Sino-Hungarian Trade Expands." Beijing Review, 28 (Dec. 30, 1985): 26-27.

781. "Nichimen Promotes Trade with China." [Japan] Beijing Review, 29 (Jan. 20, 1986): 28-29.

782. Liu, Y. "China, Japan Extend Co-Operative Ties." Beijing Review, 29 (March 3, 1986): 16-17.

783. "Nomura Expands Business in China." [Japanese investments] Beijing Review, 29 (April 14, 1986): 30-31.

784. "First Sino-Canadian Joint Venture." Beijing Review, 29 (June 9, 1986): 30.

785. Yue Haitao. "How Volkswagen Performs in China." Beijing Review, 29 (July 21, 1986): 4-5.

786. "Jardines: An Old China Hand Readies for a New Wave of Trade." Business Week (May 21, 1979): 108-9.

787. Elliott, D. "A Trickle of Trade That May Signal a Sino-Soviet Thaw." Business Week (Dec. 3, 1984): 9.

788. Jones, D.E. and Grives, R.T. "European Carmakers Start Honking at the Japanese." Business Week (April 20, 1985): 50.

789. Siggins, M. "The Big Fortune Cookie." [Canada] Canadian Business, 59 (July 1986): 16-29.

790. Noumoff, S.J. "China: The Unexplored Prospects." The Canadian Business Review, 9 (Winter 1982): 22-26.

791. Borkenau, F. "Analysis of Sino-Soviet Relations." The China Quarterly, 94 (June 1983): 345-61.

792. Howe, C. "Growth, Public Policy and Hong Kong's Economic Relationship with China." The China Quarterly, 95 (Sept. 1983): 512-33.

793. Kokubun, R. "The Politics of Foreign Economic Policy-Making in China: The Case of Plant Cancellations with Japan." The China Quarterly, 105 (March 1986): 19-44.

794. Fu, Z. and An, B. "Strengthen Our Economic and Trade Relations with Oil-Producing Countries in the Middle East." Chinese Economic Studies, 16:4 (Summer 1983): 27-36.

795. Lin, L. "Thirty Years of Sino-Japanese Trade." Chinese Eonomic Studies, 16:4 (Summer 1983): 51-62.

796. Grow, R.F. "Japanese and American Firms in China: Lessons of a New Market." Columbia Journal of World Business, 21 (Spring 1986): 49-56.

797. "China Braves the Amazon." Dun's Business Month, 124 (Aug. 1984): 22.

798. "Laying an Egg." [Japan] The Economist, 277 (Nov. 29-Dec. 5, 1980): 61-62.

799. Sandeman, H. "China's Rich Friend." [Hong Kong] The Economist, 277 (Dec. 6, 1980): 5-6+.

800. Sandeman, H. "Chinatown: A Survey of Hong Kong." The Economist, 277 (Dec. 6-12, 1980): 1-28.

801. "Volkswagen and China: The People's Car." The Economist, 285 (Dec. 11, 1982): 86.

802. "EEC Plays Its China Card." The Economist, 286 (Feb. 26-March 4, 1983): 40-41.

803. "Chile's Flexible Friend." [external debts, China] The Economist, 287 (May 14-20, 1983): 86+.

804. "Singapore and China: The Sweet Taste of Co-Operation." [jt. venture contract for offshore oil in China] The Economist, 290 (Jan. 21, 1984): 70-71.

Trade Relations, Commercial Policy, Integration 71

805. "Looking to Sweden." The Economist, 293 (Oct. 13, 1984): 84.

806. Maidment, P. "Partners Again." [Hong Kong] The Economist, 295 (May 11, 1985): 17-18.

807. "China: Beclouded by Greed." [Japan] The Economist, 296 (Aug. 10, 1985): 50+.

808. "Trade Precedes the Flag." [South Korea] The Economist, 297 (Nov. 2, 1985): 76-77.

809. Emmott, B. "East Asian Romance." [Japan] The Economist, 297 (Dec. 7, 1985): 15-16.

810. "Hong Kong Bucked by America and China." The Economist, 298 (March 1, 1986): 64+.

811. "The Hong Kong China Wants After 1997." The Economist, 299 (June 28, 1986): 33-34.

812. Hataye, J. "Japanese May Open Venture in China to Western Firms." Electronic News, 25:suppl. (Sept. 17, 1979): 26S.

813. Goodstadt, L. "Hong Kong Builds Bridges to China." Euromoney (April 1985): 153+.

814. Lee, D. "Consortia Are the Only Answer to US Strength." [Western Europe] The Far Eastern Economic Review, 103 (March 16, 1979): 93-94.

815. Bonavia, D. "New Role for the Golden Goose." [Hong Kong] The Far Eastern Economic Review, 104 (April 20, 1979): 43-44.

816. Lee, D. "China Gets an Arab Float." The Far Eastern Economic Review, 107 (March 7, 1980): 88-89.

817. Jenkins, D. "Opening the Front Door." [Indonesia] The Far Eastern Economic Review, 107 (March 14, 1980): 62-63.

818. Breithaupt, H. "Still Leery of the West's US $27 Million Credit Offer." The Far Eastern Economic Review, 109 (July 4, 1980): 59-61.

819. Subhan, M. "Allure of EEC Trade." The Far Eastern Economic Review, 109 (July 25, 1980): 43.

820. Lee, M. "Mighty City's Heavenly Hopes." [Hong Kong] The Far Eastern Economic Review, 109 (Sept. 12, 1980): 48-50.

821. Kemenade, W. van. "Double Dutch Over Two Little Subs for the KMT." [Netherlands] The Far Eastern Economic Review, 111 (Jan. 2-8, 1981): 56-59.

822. Lewis, J. and Bonavia, D. "Honeymoon Is Over." [Japan] The Far Eastern Economic Review, 111 (Feb. 20-26, 1981): 46-47.

823. Lee, M., et al. "Hong Kong '81." The Far Eastern Economic Review, 111 (March 13-19, 1981): 35-56.

824. Tanzer, A. and Smith, P. "Taiwan's China Links." The Far Eastern Economic Review, 112 (June 5-11, 1981): 53-54.

825. Curry, L. "Forced Separation." [Republic of Korea] The Far Eastern Economic Review, 117 (July 16-22, 1982): 52-53.

826. Lee, M. "Mighty Nice Gesture." [Hong Kong-China backed consortium] The Far Eastern Economic Review, 117 (Aug. 6-12, 1982): 48+.

827. Jao, Y.C. "Dependence Is a Two-Way Street." [Hong Kong] The Far Eastern Economic Review, 119 (Jan. 20, 1983): 38-42.

828. Delfs, R. "Entrepot Trade: China's Greatest Port." [Hong Kong] The Far Eastern Economic Review, 119 (March 17, 1983): 74.

829. Wisniewski, M. "Trade Way Forward." [USSR] The Far Eastern Economic Review, 119 (March 24, 1983): 80.

830. Bonavia, D. and Delfs, R. "China Goes Down Under." The Far Eastern Economic Review, 119 (March 31, 1983): 52-54.

831. Ma, T. "Way into China Oil." The Far Eastern Economic Review, 120 (June 23, 1983): 86-87.

Trade Relations, Commercial Policy, Integration   73

832. Lee, M. "A Back Seat for Politics." [USSR] The Far
      Eastern Economic Review, 127 (Jan. 10, 1985): 15-16.

833. Clad, J. "Opening Doors Policy." [Malaysia] The Far
      Eastern Economic Review, 128 (May 2, 1985): 48-49.

834. Lee, M. "Great Drive Forward." [Volkswagen/Shanghai
      Autoworks Venture] The Far Eastern Economic Review,
      128 (May 23, 1985): 74-75.

835. Islam, S. "Pact Points the Way." [Western Europe]
      The Far Eastern Economic Review, 128 (June 6, 1985):
      60-61.

836. Clad, J. "Iran Woos Peking." The Far Eastern Economic
      Review, 128 (June 13, 1985): 44-45.

837. Clad, J. "An Affair of the Head." [Malaysia] The Far
      Eastern Economic Review, 129 (July 4, 1985): 12-14.

838. Kulkarni, V.G. and Kaye, L. "Second-Hand Relations."
      [Indonesia] The Far Eastern Economic Review, 129
      (July 18, 1985): 98-99.

839. "Deutschland '85." [Germany] The Far Eastern Economic
      Review, 128 (July 20, 1985): 57-58.

840. Nations, R. "A $14 Billion Deal: Trade Agreement Eases
      Sino-Soviet Reconciliation." The Far Eastern Economic
      Review, 129 (July 25, 1985): 12-13.

841. do Rosario, L. "Economics Makes Good Bedfellows."
      [Hong Kong] The Far Eastern Economic Review, 129
      (Aug. 1, 1985): 26-29.

842. Smith, C. "China's Buy-Japanese Binge Strains the
      Relationship; A Crisis of Plenty." The Far Eastern
      Economic Review, 129 (Aug. 1, 1985): 46-47.

843. do Rosario, L. "Who Will Pick the Plum of a $1 Billion
      Market?" [Western Europe] The Far Eastern Economic
      Review, 129 (Aug. 22, 1985): 79-80+.

844. Aznam, S. "No More in the Back Seat." [Malaysia] The
      Far Eastern Economic Review, 130 (Dec. 12, 1985): 42-43.

845. Buzo, A. "Order on the Frontier." [North Korea] The
      Far Eastern Economic Review, 131 (March 20, 1986):
      136-37.

846. Smith, C. "Diagnosis: Chronic Delays; Treatment: Flexibility." [Japanese investments] The Far Eastern Economic Review, 132 (April 24, 1986): 6-7.

847. Smith, C. "The Ties That Bind." The Far Eastern Economic Review, 132 (April 24, 1986): 73-78.

848. Smith, C. "Something Ventured, Something Gained." [Japanese investments] The Far Eastern Economic Review, 132 (April 24, 1986): 78-79.

849. Delfs, R. "Offshore Motivation." [West German technical assistance] The Far Eastern Economic Review, 132 (May 22, 1986): 65-66.

850. Roscoe, B. "A Nuclear Reaction." [technical assistance, Japan] The Far Eastern Economic Review, 133 (July 3, 1986): 80-81.

851. Goldstein, C. "Growing Economic Ties But No Politcal Trust." [Taiwan] The Far Eastern Economic Review, 133 (July 24, 1986): 24-27.

852. Goldman, C. "Trade Booms Despite the Ban by Taiwan." The Far Eastern Economic Review, 133 (July 24, 1986): 29-30.

853. Emerson, R.S. "Japan's Trade with China: The Free World Stands to Gain." Finance, 96 (July 1979): 69.

854. Kraar, L. "China's Drive for Capitalist Profits in Hong Kong." Fortune, 99 (May 21, 1979): 110-12+.

855. Kraar, L. "Confidence Is Building in Hong Kong." Fortune, 109 (June 11, 1984): 140-42+.

856. Gorman, T.D. "Hong Kong/China '86: Opportunities for Growing American Businesses." Inc., 8 (May 1986): 135-36+.

857. Hansen, J. "Chinese Debut at Milan Tool Show." Industry Week, 203 (Nov. 10, 1979): 126-27.

858. Cheng, J.Y.S. "China's Japan Policy in the 1980s." International Affairs, 61 (Winter 1984/85): 91-107.

859. "Chinese Technicians Apprenticing in Japan." Journal of Commerce and Commercial, 353 (July 13, 1982): 4A.

860. "Japan Expands Vehicle Sales to China." Journal of Commerce and Commercial, 362 (Dec. 20, 1984): 1A.

861. "Closer Soviet-China Ties Spark Rail Proposals." Journal of Commerce and Commercial, 363 (Jan. 10, 1985): 1A.

862. "British Firms Flock to Chinese Marts." Journal of Commerce and Commercial, 364 (May 29, 1985): 1A.

863. Goldman, M.I. "Soviet Perceptions of Chinese Economic Reforms and the Implications for Reform in the U.S.S.R." Journal of International Affairs, 39 (Winter 1986): 41-55.

864. Jacobs, P. "Hong Kong and the Modernization of China." Journal of International Affairs, 39 (Winter 1986): 63-75.

865. Warrington, B. "Doing Business in China: Why Britons May Have the Advantage." Management Decision, 21:6 (1983): 25-30.

866. "Lure of Chinese Market Draws Japan." New York Times, 134 (March 18, 1985): D12.

867. Pollack, J.D. "China's Changing Perception of East Asian Security and Development." Orbis, 29 (Winter 1986): 771-94.

868. "Hong Kong: Cooperation with China." Petroleum Economist, 47 (Jan. 1980): 32-33.

869. "China Walks the Line Between West and Russia for Technology." [sale of military equipment] Research and Development, 28 (June 1986): 46-47.

870. Lanier, A.R. "Chinese, the Arabs: What Makes Them Buy?" Sales & Marketing Management, 122 (March 1979): 39-41.

871. "U.S. Blocking the Sale of French Goods to China." The Wall Street Journal (Aug. 15, 1985).

872. Yee, H.S. "China and the Pacific Community Concept." World Today, 39 (Feb. 1983): 68-74.

BALANCE OF PAYMENTS, INTERNATIONAL FINANCE
[430]

Balance of Payments; Mechanisms of Adjustment; Exchange Rates
[431]

873. Liu, G. "Why Is China Striving to Wipe Out Its Deficit?" Beijing Review, 24 (April 13, 1981): 21-22.

874. "China's Foreign Exchange Control Policies." Beijing Review, 28 (Oct. 28, 1985): 20-22.

875. Chen, N.-R. "People's Republic of China's Healthy Financial Status Should Facilitate Import Rise." Business America, 6 (Aug. 22, 1983): 48-49.

876. Keidel, A. and Tang, H. "Financing Modernization with a Trade Surplus." Business Week (May 30, 1983): 103-4.

877. Han, K. "On Adjustment of Foreign Trade Deficits." Chinese Economic Studies, 16:4 (Summer 1983): 78-96.

878. Wu, F.T. "External Borrowing and Foreign Aid in Post-Mao China's International Economic Policy: Data and Observations." The Columbia Journal of World Business, 19 (Fall 1984): 53-61.

879. Goodstadt, L. "Why the Renminbi Must Devalue." Euromoney (Oct. 1983): 336.

880. Lau, E. "Balance of Trade." [Hong Kong] The Far Eastern Economic Review, 128 (May 23, 1985): 92-93.

881. do Rosario, Louise. "Time to Pay the Piper." The Far Eastern Economic Review, 129 (Aug. 22, 1985): 100-1.

882. "Bankers Say China to Sell More Gold." Journal of Commerce and Commercial, 365 (July 30, 1985): 1A.

883. "McNamara Says China Becoming Debtor Nation." The Wall Street Journal (Sept. 6, 1985).

884. "China Devalues Yuan to Shore Up Economy." The Wall Street Journal (Oct. 30, 1985).

Balance of Payments, International Finance 77

International Monetary Arrangements
[432]

885. "Bank of China Seeks More Worldwide Dealings." ABA Banking Journal, 74 (Nov. 1982): 88.

886. "Bank Issues Currency, Sets Rates, Administers Planning; International Banking Handled by Ministry of Finance Thru 3 Banks." American Banker, 147 (Dec. 21, 1982): 8.

887. "An American Banker's Views of the People's Republic of China." American Banker, 147 (Dec. 21, 1982): 9.

888. "China: Back-Pedalling on Western Loans." The Banker, 129 (Sept. 1979): 21-22.

889. "First Chicago's Chinese Cracker." Banker, 130 (Sept. 1980): 73+.

890. Pastore, S.L. and Shen, P. "Banking and the China Market." The Bankers Magazine, 164 (Nov./Dec. 1981): 36-43.

891. Bleiberg, R.M. "People's Capitalism? Red China's Open-Door Policy Is Nothing to Bank On." [editorial] Barron's, 65 (Jan. 7, 1985): 9.

892. Ji, C. "China's Utilization of Foreign Funds and Relevant Policies." Beijing Review, 24 (April 20, 1981): 15-20.

893. Bu, M. "China's Financial Relations with Foreign Countries." Beijing Review, 24 (April 20, 1981): 21-23.

894. Zhao, B. "Bank of China Plays a Key Role in Open Policy." Beijing Review, 28 (June 3, 1985): 23-24.

895. "China: Foreign Bankers Make a Beeline for Beijing." Business Week (Nov. 8, 1982): 53-54.

896. "A Great Leap into Foreign Borrowing." Business Week (April 2, 1984): 43+.

897. Krafft, W. "What an American Bank Can Do for You: A Banker's Approach to China." The Columbia Journal of World Business, 14 (Summer 1979): 37-42.

898. "China Eyed by ITMF Delegates." Daily News Record, 12 (Oct. 22, 1982): 2.

899. "United, Cautious Borrower Be." The Economist, 271 (April 7, 1979): 112-13.

900. "Big Deals." [American loans] The Economist, 272 (Sept. 29, 1979): 82+.

901. "Trade with China: Thoughts for Bankers." The Economist, 273 (Nov. 10, 1979): 106.

902. "Big Borrower Cometh." [China joins IMF and World Bank] The Economist, 275 (May 17, 1980): 13.

903. "Orthodoxy Prevails; China and the IMF." The Economist, 278 (March 7-13, 1981): 70-71.

904. "China Laid Bare: Borrow and Prosper, Says the World Bank." The Economist, 279 (June 20-26, 1981): 44-45.

905. "Foreign Banks in China: Toe in a Hold." The Economist, 295 (May 4, 1985): 83-84.

906. "Banking Groups Scrambling to Help China Modernize; Obstacles to China Financing Loom." Engineering News-Record, 202 (Feb. 22, 1979): 10-11.

907. Goodstadt, L. "Peking's French Connection in Hong Kong." Euromoney (July 1979): 146-47.

908. Altschul, J. "Japanese Banks Keep Faith in China." Euromoney (July 1984): 138-39+.

909. Dahlby, T. and Saunders, I. "Japan Goes for the Bonanza." The Far Eastern Economic Review, 103 (Feb. 2, 1979): 42-44.

910. "Rate Cutting in the Euromarket." The Far Eastern Economic Review, 103 (March 2, 1979): 91-92.

911. Liu, M. "China Comes in from the Cold; Peking Enters the Euromarket." The Far Eastern Economic Review, 104 (June 1, 1979): 42-45.

912. Dahlby, T. "Japan Proceeds with Caution." The Far Eastern Economic Review, 104 (June 1, 1979): 45-46.

Balance of Payments, International Finance

913. Rowley, A. "Land of Opportunity." <u>The Far Eastern Economic Review</u>, 105 (July 6, 1979): 40-44.

914. Liu, M. "Rock-Bottom Terms Are a Must for Foreigners." <u>The Far Eastern Economic Review</u>, 105 (Sept. 21, 1979): 61-63.

915. Liu, M. "China Will Stay Cautious Despite a US $60 Billion Credit." <u>The Far Eastern Economic Review</u>, 106 (Oct. 5, 1979): 75+.

916. Dahlby, T. "Japan's China Extravaganza." <u>The Far Eastern Economic Review</u>, 106 (Nov. 16, 1979): 56-58.

917. Rowley, A. "Cementing New Relations." <u>The Far Eastern Economic Review</u>, 107 (Jan. 25, 1980): 63.

918. Bonavia, D. "No Big Loans Yet But Soon China Must Act on Options." <u>The Far Eastern Economic Review</u>, 108 (April 4, 1980): 85.

919. Gigot, P. "Banking on the Future: Look Behind the Scenes." <u>The Far Eastern Economic Review</u>, 108 (May 16, 1980): 70+.

920. "Is China a Non-Event for Bankers? Yes ... and No." <u>The Far Eastern Economic Review</u>, 109 (Sept. 19, 1980): 73-74.

921. Bonavia, D. "Need for More Foreign Finance Continues." <u>The Far Eastern Economic Review</u>, 109 (Sept. 26, 1980): 52-53+.

922. Bonavia, D. "Peking Tightens Its Belt." <u>The Far Eastern Economic Review</u>, 111 (Feb. 27, 1981): 72-73.

923. Bonavia, D. "One Back, Two Forward." <u>The Far Eastern Economic Review</u>, 112 (April 17-23, 1981): 50-51.

924. Delfs, R. "All Those Great Expectations Are Still Only a Modest Aid Flow." <u>The Far Eastern Economic Review</u>, 121 (Sept. 29, 1983): 94-95.

925. Langston, N. "The Waiting Game." <u>The Far Eastern Eonomic Review</u>, 124 (June 14, 1984): 102-3.

926. Langston, N. "Waiting--and Wishing--for Some Action." <u>The Far Eastern Economic Review</u>, 128 (April 25, 1985): 70-71.

927. Lee, M. "Foreign Bank Freedoms." The Far Eastern Economic Review, 128 (April 25, 1985): 130-31.

928. Cheng, E. "Lending Credibility." The Far Eastern Economic Review, 128 (May 9, 1985): 84-85.

929. Rowley, A. "A Tale of Three Chinas." [membership of Asian Development Bank] The Far Eastern Economic Review, 128 (May 16, 1985): 64-65.

930. "China to Receive Its First World Bank Loan." Finance and Development, 18 (Summer 1981): 2-3.

931. "The Red and the Green: Want to Startle Your Creditors?" [pay your bills with checks from your Bank of China account] Forbes, 129 (March 15, 1982): 83.

932. "World Bank to Boost China Loan." Journal of Commerce and Commercial, 359 (Jan. 6, 1984): 1A.

933. "Foreign Bank Inroads Forecast for China." Journal of Commerce and Commercial, 364 (June 14, 1985): 1A.

934. "Congressmen Seek to Block Loan to China." Textile World, 134 (July 1984): 24+.

935. Nelson, J.F. "Hong Kong's Banking Community: Hong Kong's Future Is in China's Hands." United States Banker, 96 (May 1985): 14-16+.

936. "Bank of China Arranges First Loan Outside China." The Wall Street Journal (April 24, 1985).

937. "China Poses Frustrations (As Well as Opportunities for Foreign Bankers)." The Wall Street Journal (June 11, 1985).

INTERNATIONAL INVESTMENT AND FOREIGN AID
[440]

International Investment and Capital Market
[441]

938. Gaeta, J.J. "Six Possible Sources for Financing Your Next Project in the People's Republic of China." American Import/Export Management, 100 (June 1984): 24-26.

939. "Implementation of the Income Tax Laws of the People's Republic of China: Joint Ventures with Chinese and Foreign Investment." Beijing Review, 24 (March 10, 1981): 23-25.

940. "Income Tax Law of the People's Republic of China Concerning Foreign Enterprises." Beijing Review, 24 (Dec. 28, 1981): 17-19.

941. "Detailed Rules and Regulations for the Implementation of the Income Tax Law of the People's Republic of China Concerning Foreign Enterprises." Beijing Review, 25 (April 5, 1982): 20-25.

942. Wei Yuming. "China's Policy on Absorption of Direct Investments from Foreign Countries." Beijing Review, 25 (July 26, 1982): 18-22.

943. "Another Important Document on Joint Ventures." Beijing Review, 26 (Oct. 3, 1983): 4-5.

944. "Regulations for the Implementation of the Law of the People's Republic of China on Joint Ventures Using Chinese and Foreign Investment." Beijing Review, 26 (Oct. 10, 1983): i-xvi.

945. "Investment Environment Seen as Favourable." Beijing Review, 27 (July 16, 1984): 16-19.

946. "Foreigners Entitled to Property Rights." Beijing Review, 28 (March 18, 1985): 28.

947. "Investment Crucial for Updating Industry." Beijing Review, 28 (June 3, 1985): 24-25.

948. "Foreign Economic Contract Law of the People's Republic of China." Beijing Review, 28 (July 8, 1985): i-iv.

949. "The Accounting Regulations of the People's Republic of China for the Joint Ventures Using Chinese and Foreign Investment." Beijing Review, 28 (July 8, 1985): v-xvi.

950. Jian, C. "Joint Venture: Success Speaks for Itself." Beijing Review, 28 (Nov. 4, 1985): 21-25.

951. Han, B. "Mutual Trust Is Crucial to Co-Operation." Beijing Review, 28 (Nov. 4, 1985): 25-26.

952. "Beijing Seeks Foreign Funds." Beijing Review, 28 (Nov. 4, 1985): 29-30.

953. Jing Wei. "Legal Guarantee for Foreign Investors." Beijing Review, 29 (June 2, 1986): 4-5.

954. "China Hails Entry to Asian Bank." Beijing Review, 29 (May 12, 1986): 7-8.

955. "Heilongjiang Province Attracts Foreign Capital." Beijing Review, 29 (Aug. 18, 1986): 28-29.

956. "A Foreigner's Guide to Investment in China." Beijing Review, 29 (March 17, 1986): 32-33.

957. "Foreign Firms Expand in China." Beijing Review, 28 (Nov. 11, 1985): 30-31.

958. "China Marks New Policy with Joint Venture Law to Encourage Investment." Business America, 2 (Aug. 27, 1979): 18-21.

959. Chen, N.-R. "China's New Plan Reaffirms Need for Foreign Loan and Investment." Business America, 6 (Feb. 21, 1983): 38-39.

960. "China's Bold New Program to Lure Foreign Investment." Business Week (Oct. 15, 1984): 183-84.

961. Rae, A.E.I. "Talking Business in China." The China Quarterly, 90 (June 1982): 271-80.

962. Ji, C. "Utilization of Foreign Investment and the Codification of Business Law." Chinese Economic Studies, 16:1 (Fall 1982): 53-68.

963. Yu, X. and Lin, Z. "Utilization of Foreign Capital to Renovate Enterprises Produces Good Results." Chinese Economic Studies, 16:3 (Spring 1983): 40-47.

964. Fang, S. "On the Issue of Utilizing Foreign Capital." Chinese Economic Studies, 18:4 (Summer 1985): 101-6.

965. Oster, M. "I'm Bullish on China." Commodities, 8 (April 1979): 37-39.

966. "Americans Get Japan to Help." The Economist, 270 (Jan. 20, 1979): 80-81.

967. "There's Still a Market." The Economist, 271 (June 23, 1979): 91-92.

968. "Foreign Venturers." The Economist, 273 (Dec. 29, 1979): 29-30.

969. "Investing in China." The Economist, 274 (Feb. 2, 1980): 72-73.

970. "Red Tycoons." [investment in Hong Kong by Chinese] The Economist, 277 (Dec. 6, 1980): 21-22.

971. "Foreign Devils Are Welcome Again." The Economist, 283 (April 3-9, 1982): 87-89.

972. "China's Lopsided Investment." The Economist, 290 (Jan. 21, 1984): 71.

973. "Markets and Mao Mix Madly." The Economist, 290 (Jan. 28, 1984): 14.

974. "Hong Kong and Shanghai Banking Corporation Disciplining Hong Kong's Banks." The Economist, 296 (July 13, 1985): 84-85.

975. Liu, M. "China Keeps Its Project Partners Guessing; Straightening Out the Guidelines." The Far Eastern Economic Review, 102 (Dec. 15, 1978): 84-87.

976. Liu, M. "Loyalty and Rock-Bottom Terms." The Far Eastern Economic Review, 104 (June 22, 1979): 98-100.

977. Liu, M. "Reincarnation of Investment Companies." The Far Eastern Economic Review, 105 (July 27, 1979): 70-71.

978. Liu, M. "China Puts Hong Kong Investors at Ease." The Far Eastern Economic Review, 104 (Aug. 20, 1979): 42-43.

979. Loong, P. "In Peking We Trust." The Far Eastern Economic Review, 108 (May 30, 1980): 79-80.

980. Lee, M. "Leasing Helps in Greasing Business." The Far Eastern Economic Review, 108 (June 6, 1980): 64.

981. Srodes, J. "Rockefeller Meets Rong." The Far Eastern Economic Review, 108 (June 13, 1980): 108-10.

982. Lewis, J. and Rowley, A. "Battle of Baoshan." [Japanese investments] The Far Eastern Economic Review, 11 (Feb. 20-26, 1981): 49-50+.

983. Kurata, P. "Lower Costs Lure China Traders." The Far Eastern Economic Review, 111 (Feb. 27, 1981): 69-70+.

984. Lee, M. "New Style Compradores for Chinese Foot-Dragging." The Far Eastern Economic Review, 111 (March 13-19, 1981): 47+.

985. Lee, M. "Golden Age of Our Economic Relations." The Far Eastern Economic Review, 111 (March 13-19, 1981): 49-50+.

986. Loong, P. "Foreign Developers Lose Their Heady Optimism: Counting the Cost After Two Wasted Years of Talks." The Far Eastern Economic Review, 115 (Feb. 26-March 4, 1982): 90-91.

987. Brown, B. "It's Being So Conservative That Makes China Attractive." The Far Eastern Economic Review, 124 (April 26, 1984): 95-96.

988. Frank, R. "The Outlook for Foreigners Is Now Slightly Rosier." The Far Eastern Economic Review, 124 (April 26, 1984): 94-95.

989. Bonavia, D. "Investing in the Old Country." The Far Eastern Economic Review, 126 (Nov. 22, 1984): 50-51.

990. Langston, N. "Laying Down the Law." The Far Eastern Economic Review, 127 (Jan. 24, 1985): 64-65.

International Investment and Foreign Aid

991. "The Curtain Goes Up." *The Far Eastern Economic Review*, 127 (Jan. 31, 1985): 50-51.

992. Hanson, R. "Funding the Future: China Taps the Tokyo Bond Market." *The Far Eastern Economic Review*, 127 (Feb. 7, 1985): 90-91.

993. Delfs, R. "Changing the Pattern." *The Far Eastern Economic Review*, 128 (May 9, 1985): 70-71.

994. Langston, Nancy. "Railroaded Euromarket (The $64,000+ Question)." *The Far Eastern Economic Review*, 128 (May 23, 1985): 70-71.

995. Bowring, P. "CITIC: Another Domino Wobbles." *The Far Eastern Economic Review*, 131 (Jan. 23, 1986): 50-53.

996. Lee, M. "New Rules Help, But It's Still a Struggle for Some." *The Far Eastern Economic Review*, 131 (March 20, 1986): 87-89.

997. Nevans, A. "China: What's in It for Investors?" *Financial World*, 148 (Feb. 1, 1979): 14-17.

998. "China's Frozen Assets." *Financial World*, 148 (April 15, 1979): 30-31.

999. "China Mulls Revival of Stocks and Bonds." *Journal of Commerce and Commercial*, 360 (June 8, 1984): 1A.

1000. "China Launches Foreign Investment Drive." *Journal of Commerce and Commercial*, 362 (Nov. 7, 1984): 5A.

1001. Huan, G. "China's Open Door Policy, 1978-1984." [foreign investments] *Journal of International Affairs*, 39 (Winter 1986): 1-18.

1002. Hammer, A. "On a Vast China Market." *Journal of International Affairs*, 39 (Winter 1986): 19-25.

1003. Rong, Y. "China's Open Policy and CITIC's Role." *Journal of International Affairs*, 39 (Winter 1986): 57-61.

1004. "The New Legal Framework for Joint Ventures in China: Guidelines for Investors." *Law and Policy in International Business*, 16 (Fall 1984): 1005-50.

1005. Wu, F.W. "Political Risk of Foreign Direct Investment in Post-Mao China: A Preliminary Assessment." Management International Review, 22:1 (1982): 13-25.

1006. "Brokering Deals Between East and West." New York Times, 134 (Jan. 13, 1985): F6.

1007. Wu, F. "Realities Confronting China's Foreign Investment Policy." World Economics, 7:3 (Sept. 1984): 295-311.

International Business
[442]

1008. Thomas, P.G. "Humanity Softens Ritualized Edges in China." Advertising Age, 54 (April 25, 1983): sec. 2: M16.

1009. "McDonnell Will Make Airlines with Chinese." American Metal Market, 92 (Feb. 6, 1984): 1.

1010. "Signal Finds China Concentrating on Basic Equipment." American Metal Market, 93 (Sept. 9, 1985): 4.

1011. "China Boosts Efforts to Acquire Technology, Aerospace Equipment." Aviation Week, 122 (May 27, 1985): 16.

1012. "Foreign Trade Targets Overfulfilled." Beijing Review, 24 (Feb. 2, 1981): 6-7.

1013. "Machine Export and Import Increase." Beijing Review, 28 (Jan. 14, 1985): 29-30.

1014. "Fujian Plans for More Imports." Beijing Review, 28 (June 17, 1985): 32-33.

1015. "China Exports More Medicine." Beijing Review, 29 (March 3, 1986): 28-29.

1016. "Phoenix in the East." [Insurance in China] Best's Review. Property-Casualty Insurance Edition, 84 (Nov. 1983): 40.

1017. Mullins, P.E. "Coke's First China Plant--Designed with Market Operation in Mind." Beverage Industry, 71 (Nov. 20, 1981): 112-14.

International Investment and Foreign Aid 87

1018. Toy, C.D. "People's Republic of China: New Regulations on Foreign Exchange Balancing in Joint Ventures." Bulletin for International Fiscal Documentation, 40:4-5 (April/May 1986): 154-56.

1019. Moore, T.E. "Analysis of the Chinese Market for Light Industry." Business America, 5 (May 3, 1982): 16-19.

1020. Matheson, J. "Doing Business with China." Business America, 8 (June 10, 1985): 5-6.

1021. "Motor Vehicle Production in China Lures Automakers." Business Japan, 29 (July 1984): 16-18+.

1022. Bohn, J. "AMC Launches Attack on Japan from the Rear." Business Marketing, 68 (Sept. 1983): 24+.

1023. "China's Low-Cost Labor Has Rivals Sweating." Business Week (May 17, 1982): 40.

1024. "A Seattle Steel Plant Heads for a New Life in Shanghai." Business Week (Dec. 12, 1983): 32-33.

1025. "A New Burst of Business Enthusiasm over China." Business Week (April 30, 1984): 37-38.

1026. Tarpey, J.P. "WJS Inc.: Cutting Deals in the New China." Business Week (July 22, 1985): 88.

1027. Nehemkis, P. and Nehemkis, A. "China's Law on Joint Ventures." California Management Review, 22:4 (Summer 1980): 37-46.

1028. "Rights, Wrongs and Rituals." [Chinese business etiquette and approach to doing business] Canadian Business, 59 (July 1986): 22-23.

1029. Garcia-Borras, T. "China and the Chemical Engineer." Chemical Engineering (July 22, 1985): 65-68.

1030. "Kellogg Hopes to Form Joint Venture Company for Products in China." Chemical Marketing Reporter, 226 (Oct. 22, 1984): 5+.

1031. Gibson, W.D. "China's Door Is Open But It Takes Time to Get In." Chemical Week, 124 (Jan. 24, 1979): 32-34.

1032. "China: A Billion-Dollar Chemical Market." *Chemical Week*, 131 (Sept. 29, 1982): 55-56.

1033. "China Raises the Bamboo Curtain." *Chemical Week*, 135 (Aug. 22, 1984): 64-67.

1034. "China's Lure for Kellogg Rust." *Chemical Week*, 135 (Dec. 5, 1984): 14+.

1035. Block, P.M. and Watzmann, A. "PPG Nails Down China Ventures." *Chemical Week*, 136 (May 1, 1985): 17.

1036. "China's Modern Spice Trade." *China Business Review*, 9 (May/June 1982): 8-16.

1037. "Getting a Shoe in the Door." *China Business Review*, 9 (May/June 1982): 40-44.

1038. "China's Door Open Wider for Business." *Colorado Business*, 12 (June 1985): 20.

1039. Tsurumi, Y. "Two Models of Corporation and International Transfer of Technology." *The Columbia Journal of World Business*, 14 (Summer 1979): 43-50.

1040. "Chinese Contracts: Foreign Devils Are Welcome Again." *The Economist*, 283 (April 3, 1982): 87-88.

1041. "IBM: Fast Track to China." *The Economist*, 293 (Nov. 17, 1984): 68+.

1042. "The Lure of China Once More Entices Merchant Kings." *EDP Industry Report*, 20 (April 23, 1985): 6.

1043. Furst, A. "U.S. T&M Makers Scale Great Wall of China." [test & measurement industries] *Electronic Business*, 11 (March 15, 1985): 28-29.

1044. Ehrlich, P. "China Plan to Draw Firms Often Results in Headaches." *Electronic News*, 29:suppl. (Oct. 10, 1983): 6+.

1045. "Patience Wins China Deals, Says ITT's Jenish." *Electronics*, 58 (Sept. 16, 1985): 40.

1046. "Sales to China Are a Bright Spot for Computer Business." *Electronics Week*, 58 (May 20, 1985): 32.

1047. "Pouring into the China Set." The Far Eastern Economic Review, 103 (Jan. 26, 1979): 50-51.

1048. Maxwell, B. "International Force in the Making--Soviet Style?" [merchant marine] The Far Eastern Economic Review, 111 (Feb. 6-12, 1981): 45-46.

1049. "China as Mass Market Appears to Be a Dream." Footwear News, 39 (April 25, 1983): S2.

1050. "China Still Not Importer's Paradise." Footwear News, 48 (May 6, 1985): 4.

1051. Tanzer, A. "Comrade Innkeeper (F. Hsieh)." Forbes, 136 (July 1, 1985): 120-21.

1052. Gilbert, N. "The China Guanxi." [lack of case law protection for joint ventures] Forbes, 136 (July 29, 1985): 104.

1053. Kraar, L. "China after Marx: Open for Business?" Fortune, 111 (Feb. 18, 1985): 28-33.

1054. Roby, J.L. "Is the China Market for You?" Harvard Business Review, 58 (Jan./Feb. 1980): 150-51+.

1055. "International Business: Capitalism in China." Industry Week, 223 (Nov. 12, 1984): 19-20.

1056. "Techniques for Exporting to China." International Trade Forum, 19 (Jan./March 1983): 8.

1057. "Chinese Will Buy and Assemble 25 DC-9 Super 80 Airplanes." Iron Age, 227 (Feb. 6, 1984): 15.

1058. Valigra, L. "When Doing Business in China, Expect the Unexpected." Mini-Micro Systems, 18 (March 1985): 57-58.

1059. "China: A Slumbering Giant Coming Awake." Modern Machine Shop, 55 (March 1983): 84; (April 1983): 88.

1060. Tuthill, M. "China: Open for Business?" Nation's Business, 67 (Feb. 1979): 28-32+.

1061. "China's Passion for the Computer: The World's Largest Untapped Market Is at Stake." New York Times, 134 (Jan. 6, 1985): F1.

1062. "U.S. Small Business Is Drawn to China; Trips Pay Off in Contracts." New York Times, 134 (Sept. 9, 1985): D6.

1063. Hart, P. "China Hikes Imports of Advanced Energy Technology, Equipment." Oil & Gas Journal, 82 (Aug. 13, 1984): 41-46.

1064. "Imports Boost China Petrochemicals." Oil & Gas Journal, 82 (Sept. 10, 1984): 80-81.

1065. "China Opportunity: Amateurs Need Not Apply." Personnel Administrator, 30 (July 1985): 97.

1066. Denny, D.L. "Doing Business with China." Problems of Communism, 30 (July/Aug. 1981): 69-80.

1067. "U.S. Allies Speed Technology Export to China." Research and Development, 28 (Jan. 1986): 40-41.

1068. "Industrial Marketing: Cracking the China Market." Sales and Marketing Management, 125 (July 7, 1980): 9-10.

1069. Israel, E.A. "Bloomingdale's Embraces China." Stores, 62 (Aug. 1980): 41-43.

1070. Bennett, A. "Western Companies Find They Must Change Their Strategies as China Eases Government Out of Business Transactions." The Wall Street Journal (April 3, 1985).

1071. Bennett, A. "Firm Finds There Aren't Any Shortcuts in China." The Wall Street Journal (April 25, 1985).

1072. "Your Friendly Chinese Small-Arms Merchant." The Wall Street Journal (June 17, 1985).

1073. Smith, D.C. "Chrysler Talks Joint Venture with Chinese." Ward's Auto World, 21 (May 1985): 44.

International Aid
[443]

1074. "Aid Precedes Trade." The Economist 273 (Dec. 8, 1979): 71-72.

1075. Delfs, R. "All Those Great Expectations Are Still Only a Modest Aid Flow." The Far Eastern Economic Review, 121 (Sept. 29, 1983): 94-95.

ADMINISTRATION, BUSINESS FINANCE, MARKETING, ACCOUNTING
[500]

ADMINISTRATION
[510]

1076. "Bankruptcy Law Under Discussion." Beijing Review, 28 (July 7, 1986): 7-9.

1077. "First Factory to Go Bankrupt." Beijing Review, 29 (Aug. 18, 1986): 5-6.

1078. Manion, M. "The Cadre Management System, Post-Mao: The Appointment, Promotion, Transfer and Removal of Party and State Leaders." The China Quarterly, 102 (June 1985): 203-33.

Enterprise Organization and Decision Theory
[511]

1079. Johnson, D.G. "Cracks in the Iron Rice Bowl." Across the Board, 21 (Feb. 1984): 34-41.

1080. Tian, Y. "More Authority for Enterprises Revives the Economy." Beijing Review, 24 (April 6, 1981): 21-25.

1081. "Enterprises Reform the Cadre System." Beijing Review, 26 (Jan. 31, 1983): 5-6.

1082. "Reforming Enterprise Leadership System." Beijing Review, 27 (June 18, 1984): 4-5.

1083. Wong, S.-L. "The Chinese Family Firm: A Model." The British Journal of Sociology, 36 (March 1985): 58-72.

1084. Yu, G. "Forecasts, Management, and Long-Range Considerations." Chinese Economic Studies, 16:1 (Fall 1982): 30-40.

1085. Reeder, J.A. "Entrepreneurship in the People's Republic of China." The Columbia Journal of World Business, 19 (Fall 1984): 43-51.

1086. "Free Enterprise Revolution." The Economist, 286 (March 5-11, 1983): 38.

1087. Loong, P. "In Peking We Trust." The Far Eastern Economic Review, 108 (May 30, 1980): 79-80.

1088. Laaksonen, O. "Participation Down and Up the Line: Comparative Industrial Democracy Trends in China and Europe." International Social Science Journal, 36:2 (1984): 299-318.

Managerial Economics
[512]

1089. Tung, R.L. "Patterns of Motivation in Chinese Industrial Enterprises." The Academy of Management Review, 6 (July 1981): 481-89.

1090. Bucknall, K.B. "Some Evidence on Managerial Behavior in Chinese Industry." ACES Bulletin, 21:2 (Summer 1979): 42-52.

1091. "The Management of China's Modernization and Its Impact on the Rest of the World." Australian Journal of Management, 7 (June 1982): 1-8.

1092. "Chinese Push Managerial Improvement." Aviation Week, 118 (June 20, 1983): 68.

1093. Ye Yinsong. "Review of Thirty Years of Management of Commercial Undertakings and Suggestions for Its Future Reform." Chinese Economic Studies, 14:4 (Summer 1981): 3-16.

1094. Yu, G. "On the Reform of the Economic Management System." Chinese Economic Studies, 17:1 (Fall 1983): 65-73.

1095. "Improve the Leadership's Work Style." Beijing Review, 25 (June 14, 1982): 7-8.

1096. "Full Responsibility for Factory Heads." Beijing Review, 27 (Aug. 6, 1984): 6-7.

1097. Segal, G. "Goodbye to Maoist Management?" The Director, 32 (Jan. 1980): 21-22.

1098. "Who Manages What." The Economist, 273 (Dec. 29, 1979): 26-28.

1099. Beedham, B., et al. "Enterprising State." The Economist, 273 (Dec. 29, 1979): 26-29.

1100. Bonavia, D. "Of Mothers-in-Law and Hairy Crabs." [Industrial management] The Far Eastern Economic Review, 103 (March 23, 1979): 21-23.

1101. Delfs, R. "More Stick, Less Carrot." The Far Eastern Economic Review, 121 (July 21, 1983): 48-49.

1102. Lasdon, L.S. "Operations Research in China." Interfaces, 10 (Feb. 1980): 23-27.

1103. "The Role of 'Face' in the Organizational Perceptions of Chinese Managers." International Studies of Management and Organization, 13 (Fall 1983): 92-123.

1104. Gittings, J. "Wages and Management in China." Journal of Contemporary Asia, 9:1 (1979): 53-66.

1105. Redding, S.G. "Cognition as an Aspect of Culture and Its Relations to Management Processes: An Exploratory View of the Chinese Case." The Journal of Management Studies, 17 (May 1980): 127-48.

1106. Vander Weele, R. "Catching China Fever: A Management Accountant's Perspective." Management Accounting, 67 (Oct. 1985): 19-25+.

1107. Chastain, C.E. "Management: The Key to China's Development." Management International Review, 22:1 (1982): 5-12.

1108. Braddick, B. and Foy, N. "Management After Mao." Management Today (Aug. 1980): 46-49+.

1109. Lowe, L. "China's Managerial Revolution." Management Today (March 1983): 68-73.

1110. Hartzell, B. "Why Study Chinese Personnel Management?" Personnel Journal, 61 (Oct. 1982): 724.

1111. Fischer, W.A. "Do We Stand on Our Heads While We Work?" [Chinese research & development managers] Research Management, 26 (March-April 1983): 28-33.

1112. Lockett, M. and Littler, C.R. "Trends in Chinese Enterprise Management, 1978-1982." World Development, 11:8 (Aug. 1983): 683-704.

Business and Public Administration
[513]

1113. "Chinese Take 'Free' Out of Enterprise." New York Times, 133 (Sept. 15, 1984): 2.

Goals and Objectives of Firms
[514]

1114. He, J. "Expansion of the Enterprise's Decision-Making Power and Change in the Ownership Relation." Chinese Economic Studies, 19:1 (Fall 1985): 10-16.

1115. "Breach of Faith." The Economist, 277 (Nov. 1-7, 1980): 83.

BUSINESS FINANCE AND INVESTMENT
[520]

1116. Ham Guang. "Capital Construction: Achievements and Problems." Beijing Review, 25 (March 29, 1982): 17-20.

1117. Wei, L. "Investors Encouraged to Aid Economy." Beijing Review, 28 (June 24, 1985): 18-19.

1118. "Investing in China." The Economist, 274 (Feb. 2, 1980): 72-73.

1119. do Rosario, L. "Bumpy Capitalist Road [Millie's Group]." The Far Eastern Economic Review, 127 (Jan. 31, 1985): 72-74.

1120. "China Venturing Down the Capitalist Road." Institutional Investor, 18 (Dec. 1984): 290.

MARKETING AND ADVERTISING
[530]

1121. Unna, W. "Advertising: China's Link Between Communistic and Capitalistic Ideas?" Advertising Age, 50 (June 11, 1979): sec. 2: S9.

1122. "Advertising in China [Special Report]." Advertising Age, 50 (Aug. 20, 1979): sec. 2: S12-S14.

1123. "Selling to China? Recognize Your Audience First." Advertising Age, 50 (Dec. 10, 1979): sec. 2: S10.

1124. Cohen, S.E. "Advertising in China." Advertising Age, 51 (Sept. 8, 1980): 43-44.

1125. Chase, D. "Is Consumer Boom a Fading Dream? [Special Report]." Advertising Age, 52 (Dec. 14, 1981): sec. 2: S1-S11.

1126. "Are You Still High on China? [Panel Discussion]." Advertising Age, 52 (Dec. 14, 1981): sec. 2: S2.

1127. Burstein, D. "Consumers Offer Critique of Ads, Consumerism." Advertising Age, 52 (Dec. 14, 1981): sec. 2: S4.

1128. Matsuda, M. "Dentsu Eases Through Open Door." Advertising Age, 52 (Dec. 14, 1981): sec. 2: S9.

1129. Gage, T.J. "Pipeline Image Ads to China." Advertising Age, 53 (May 17, 1982): sec. 2: M20-M21.

1130. Curry, L. "China Likely to Curb Foreign Products' Ads." Advertising Age, 53 (July 12, 1982): 24.

1131. Chase, D. "U.S. Stumbles at China's Wall: Aussies Filling Ad Niche." Advertising Age, 55 (May 3, 1984): 1.

96   Administration, Business Finance, Marketing, Accounting

1132. Passow, S. "Eye on Advertising in the People's Republic of China; Patience Will Pay Off." Advertising World (April-May 1983): 28+.

1133. "Exclusive: Amoco Builds Awareness in China." Advertising World (April/May 1983): 30+.

1134. "One Billion Customers?" Asian Affairs, 16 (Oct. 1985): 265-72.

1135. "Concerning Advertisements." Beijing Review, 24 (May 18, 1981): 3-4.

1136. "Socialist China's Advertising." Beijing Review, 25 (June 7, 1982): 6-7.

1137. "False Advertising Angers Consumers." Beijing Review, 28 (July 29, 1985): 8-9.

1138. "Doing Business in China: Initiating a Marketing Campaign." Business America, 3 (Sept. 8, 1980): 7-9.

1139. "Billings Are Up in China, of All Places." Business Week (June 4, 1984): 36.

1140. "China's Fledgling Advertising Industry." China Business Review, 11 (Jan./Feb. 1984): 12-17.

1141. Lee, M. "See Now, Buy Later." The Far Eastern Economic Review, 104 (May 4, 1979): 81-82.

1142. Liu, M. "Coping with the Marlboro Man." The Far Eastern Economic Review, 106 (Dec. 14, 1979): 15-16.

1143. Donath, B. "Peddling to the People's Republic of China." Industrial Marketing, 65 (Nov. 1980): 74-75.

1144. "Selling China: Guidelines." Industrial Marketing Management, 64 (Nov. 1979): 74-75.

1145. Anderson, M.H. "China's Great Leap Toward Madison Avenue." Journal of Communication, 31 (Winter 1981): 10-22.

1146. Reid, B.C. "A Tourist's Perceptions of Marketing in the People's Republic of China." Marketing & Media Decisions, 17 (Oct. 1982): 94+.

ACCOUNTING
[540]

1147. Bonavia, D. "Red Ink on China's Ledger." *The Far Eastern Economic Review*, 109 (Sept. 5, 1980): 69.

1148. Lappen, A.A. "Back to the Abacus?" *Forbes*, 128 (Aug. 17, 1981): 78.

INDUSTRIAL ORGANIZATION, TECHNOLOGICAL CHANGE, INDUSTRIAL STUDIES [600]

INDUSTRIAL ORGANIZATION AND PUBLIC POLICY [610]

1149. Lieberthal, K. "China Faces the Fundamental Revolution." Asia, 1 (May 1978): 2-9.

1150. Halpern, N.P. "China's Industrial Economic Reforms: The Question of Strategy." Asian Survey, 35 (Oct. 1985): 998-1012.

1151. "Good Start in Enlivening Key Enterprises." Beijing Review, 28 (Dec. 2, 1985): 4-5.

1152. Chan, W.K.K. "The Organizational Structure of the Traditional Chinese Firm and Its Modern Reform." Business History Review, 56:2 (Summer 1982): 218-35.

1153. "Industry's Promise." The Economist, 297 (Dec. 21, 1985): 67-69.

1154. Liu, M. "Painting a New Portrait." The Far Eastern Economic Review, 105 (July 6, 1979): 36-37.

1155. Delfs, R. "China Off the Boil." The Far Eastern Economic Review, 132 (June 5, 1986): 56-57.

1156. Solinger, D.J. "Industrial Reform: Decentralization, Differentiation, and the Difficulties." Journal of International Affairs, 39 (Winter 1986): 105-18.

Market Structure: Industrial Organization
and Corporate Strategy
[611]

1157. "Restructuring of Industry." Beijing Review, 24 (April 27, 1981): 5-6.

1158. "Individual Industry and Commerce." Beijing Review, 26 (April 25, 1983): 4-5.

1159. Liu, M. "Rival Form of Comradeship." The Far Eastern Economic Review, 106 (Oct. 12, 1979): 40-42.

Public Policy Towards
Monopoly and Competition
[612]

1160. Liu, M. "Toning Down the Spirit." [corporate law] The Far Eastern Economic Review, 105 (July 27, 1979): 83-84.

1161. Conway, M. "Will Mainland China Emerge as a New Hub of Private Enterprise?" Industrial Development, 153 (March-April 1984): 4-8.

Public Enterprises
[613]

1162. Byrd, W. "Enterprise-Level Reforms in Chinese State-Owned Enterprises." The American Economic Review, 73 (May 1983): 329-32.

1163. Byrd, W. "Enterprise-Level Reforms in Chinese State-Owned Industry." American Economic Review, 73:2 (May 1983): 329-32.

1164. Jiyun, T. "Problems in the 'Substitution of Taxes for Profits' in State-Run Enterprises." Chinese Economic Studies, 17:2 (Winter 1983/84): 68-77.

1165. Delfs, R. "Collective Efforts Are Overwhelming State Enterprise." The Far Eastern Economic Review, 131 (March 20, 1986): 70+.

Industrial Organization and Public Policy 101

1166. Kraus, R. "Bureaucratic Privilege as an Issue in Chinese Politics." World Development, 11:8 (Aug. 1983): 673-82.

Transportation
[614]

1167. Li Haibo. "China's Railways: Its Role in the Modernization Drive." Beijing Review, 27 (July 23, 1984): 25-30.

1168. Wang, D. "Developing Road and River Shipping." Beijing Review, 28 (July 15, 1985): 4-5.

1169. "Government to Aid Rural Road Repair." Beijing Review, 28 (July 29, 1985): 7-8.

1170. "China Ready to Improve Transport." Beijing Review, 28 (Oct. 21, 1985): 9-10.

1171. Guo, H. "Current Status of China's Transport and Prospects in the Near Future." Chinese Economic Studies, 17:4 (Summer 1984): 63-71.

1172. Lauriat, G. "Ro-Ro Steam Ahead to Maritime Power Status." The Far Eastern Economic Review, 103 (Feb. 9, 1979): 42-43.

1173. Lauriat, G. "Determining the Pace of China's Modernisation." The Far Eastern Economic Review, 107 (March 7, 1980): 68-69.

1174. Westlake, M. "General Administration of Civil Aviation of China; Change of Direction." The Far Eastern Economic Review, 128 (June 13, 1985): 118-21.

1175. Lyons, T.P. "China's Cellular Economy: A Test of the Fragmentation Hypothesis." Journal of Comparative Economics, 9 (June 1985): 125-44.

1176. Lyons, T.P. "Transportation in Chinese Development, 1952-1982." Journal of Developing Areas, 19:3 (April 1985): 305-28.

1177. Hanham, R.Q. and Chang, H.Y. "Scalar Variation and Nodal Accessibility in the Chinese Railroad Network." Professional Geographer, 31 (Nov. 1979): 388-92.

1178. "Chinese Take-Out?" Railway Age, 184 (Sept. 1983): 105-6.

TECHNOLOGICAL CHANGE, INNOVATION, RESEARCH AND DEVELOPMENT
[620]

1179. Baark, E. "China's Technological Economics." Asian Survey, 21 (Sept. 1981): 977-99.

1180. "Changes in Light and Heavy Industries' Proportion." Beijing Review, 24 (Jan. 26, 1981): 7-8.

1181. "New Guideline for Science and Technology." Beijing Review, 24 (April 20, 1981): 6-7.

1182. "Importing Technology--New Pattern." Beijing Review, 24 (July 13, 1981): 5-6.

1183. "China's Technical Structure." Beijing Review, 24 (Aug. 3, 1981): 22-23.

1184. "Shanghai Leads in Modernization March." Beijing Review, 25 (Jan. 4, 1982): 19-27.

1185. "Shanghai Industry in Adjustment: Blazing a New Trail." Beijing Review, 25 (Jan. 18, 1982): 18-25.

1186. "Chinese-Type Modernization: Two Keys to Industrial Development." Beijing Review, 26 (Jan. 31, 1983): 16-20.

1187. "Expand Industry Through Technical Transformation." Beijing Review, 26 (Feb. 28, 1983): 21-25.

1188. "New System Improves Industry." Beijing Review, 26 (April 11, 1983): 24-25.

1189. "Making Science Serve Economy." Beijing Review, 27 (March 5, 1984): 10-11.

1190. Lu Dong. "China's Industry on the Upswing." Beijing Review, 27 (Aug. 27, 1984): 18-21.

1191. "The Way to Vitalize Enterprises." Beijing Review, 27 (Nov. 5, 1984): 4-5.

1192. Jin, Q. "Controlling Industrial Development." Beijing Review, 28 (July 29, 1985): 4-5.

1193. Yang, D.J. "The Venture Communists Setting Up Shop in China." Business Week (July 7, 1986): 70.

1194. Walder, A.G. "China Turns to Industry Reform." Challenge, 28:1 (March/April 1985): 42-47.

1195. Conroy, R. "Technological Innovation in China's Recent Industrialization." The China Quarterly, 97 (March 1984): 1-23.

1196. Gu, N. "China's Current Efforts to Import Technology and Its Prospects." Chinese Economic Studies, 14:1 (Fall 1980): 54-67.

1197. Zhou, S. "New Advances in Science and Technology and Economic Management." Chinese Economic Studies, 19:1 (Fall 1985): 17-25.

1198. "Can the People's Republic Catch Up?" Computerworld, 17 (Nov. 14, 1983): ID 15.

1199. "China's Goal to Build Own Computer Industry." Computerworld, 17 (Nov. 14, 1983): ID 25.

1200. Hung-Ying, C. "People's Republic of China's Modernizations Depend on Import of Technology." Data Management, 18 (July 1980): 35-37.

1201. Bhalla, A.S. "Technological Transformation in China." Economia Internazionale, 37:1-2 (Feb./May 1984): 4-19.

1202. Bonavia, D. "Most Momentous Reconciliation Ever." The Far Eastern Economic Review, 104 (March 16, 1979): 48-50.

1203. Bonavia, D. "Another Turnaround in China." The Far Eastern Economic Review, 110 (Dec. 12-18, 1981): 60+.

1204. Delfs, R. "Revolution Run Riot." The Far Eastern Economic Review, 129 (July 11, 1985): 54-55.

1205. Tow, W.T. "Science and Technology in China's Defense." Problems of Communism, 34 (July/Aug. 1985): 15-31; (Nov./Dec. 1985): 87-88.

1206. Agres, T. "Technology and China." [special report] Research & Development, 27 (Feb. 1985): 119-30.

1207. "Modernization Forces Review of China's Science Institutions." Research & Development, 27 (Feb. 1985): 126+.

INDUSTRY STUDIES
[630]

1208. Maier, J.H. "Information Technology in China." Asian Survey, 20 (Aug. 1980): 860-75.

1209. Barbour, N.S. "China Revamps a Haphazard Industry." Automotive Industries, 162 (Jan. 1982): 56-57.

1210. "Newly Built Industrial Projects." Beijing Review, 24 (March 9, 1981): 5-6.

1211. "Large Petrochemical Fibre Plants Completed." Beijing Review, 24 (Sept. 21, 1981): 5-6.

1212. "Industrial Situation." Beijing Review, 24 (Sept. 28, 1981): 5-6.

1213. "General Survey of China's Industry." Beijing Review, 26 (April 18, 1983): 24-25.

1214. "China's Driving Light Industry." Beijing Review, 27 (June 4, 1984): 24-29.

1215. "Service Trades on the Rise." Beijing Review, 28 (Jan. 28, 1985): 4-5.

1216. "Auto Industry Hits Boom Times." Beijing Review, 28 (March 11, 1985): 6-8.

1217. Han, B. "Economic Reforms Revitalize Cotton Mill." Beijing Review, 28 (Oct. 14, 1985): 16-20.

Industry Studies

1218. Wang, A. "Aeronautics Industry Takes Off." Beijing Review, 29 (Jan. 7, 1986): 20-27.

1219. "China: A Great Leap Forward in Merchant Shipping." Business Week (April 2, 1979): 40+.

1220. "Mismanagement Puts Major Projects in Peril." Business Week (Sept. 8, 1980): 53.

1221. "China Slows to a Dragon's Pace." Chemical Week, 125 (July 25, 1979): 37-38.

1222. Field, R.M. "Changes in Chinese Industry Since 1978." The China Quarterly, 100 (Dec. 1984): 742-61.

1223. Davis, D.A. "China Accelerates Cosmetic Build-Up." Drug & Cosmetic Industry, 133 (Aug. 1983): 32-34.

1224. Schwartz, L. "Doubts China Electronic Industry Will Reach World Standards, Goal." Electronic News, 29: suppl. (Aug. 15, 1983).

1225. Bonavia, D. "Heavy Industry Resurgent--Leaner, More Efficient." The Far Eastern Economic Review, 109 (Sept. 26, 1980): 70+.

1226. Tharp, M. "Out of the Melting Pot." The Far Eastern Economic Review, 120 (May 5, 1983): 128-29+.

1227. Delfs, R. "Return to the Centre." The Far Eastern Economic Review, 120 (June 23, 1983): 56-58.

1228. Disney, R. "Scale and Efficiency in the Chinese Nitrogenous Fertilizer Industry." Indian Economic Review, 14:2 (Oct. 1979): 147-62.

1229. Chu, W.W. "Report from Mainland China." Textile Industries, 146 (Nov. 1982): 42-44+.

1230. Suchecki, S.M. "China the Joker in the Deck." Textile Industries, 147 (March 1983): 38-40.

1231. Peter, L.J. "China's Textile Market." Textile Industries, 147 (March 1983): 42-43+.

ECONOMIC CAPACITY
[640]

1232. Harrison, M. "Investment Mobilization and Capacity Completion in the Chinese and Soviet Economies." Economics of Planning, 19:2 (1985): 56-75.

1233. Hake, B. "Has the Time Come to Start Manufacturing in China?" Planning Review, 14 (May 1986): 44-46.

AGRICULTURE; NATURAL RESOURCES
[700]

AGRICULTURE
[710]

1234. Dernberger, R.F. "Agricultural Development: The Key Link in China's Four Modernizations Program." American Journal of Agricultural Economics, 62:2 (May 1980): 331-38.

1235. "Chinese Agriculture: Development, Production, and Trade." American Journal of Agricultural Economics, 62 (May 1980): 331-61.

1236. Lardy, N. "Chinese Agriculture: Development, Production and Trade." American Journal of Agricultural Economics, 62:2 (May 1980): 356-58.

1237. Kilpatrick, J.A. "Chinese Agriculture: Development, Production, and Trade." American Journal of Agricultural Economics, 62:2 (May 1980): 359-61.

1238. "Agriculture in the Soviet Union and China: Implications for Trade." American Journal of Agricultural Economics, 67 (Dec. 1985): 1044-66.

1239. "Two Former Models Try New Contracts." Beijing Review, 27 (Nov. 12, 1984): 8-9.

1240. "Incentives That Reward China Farmers Stimulate Record Crops." Chicago Tribune (Jan. 27, 1985): sec. 18: 15.

1241. Yu Guangyuan. "Strengthen Research on the Problem of the Economics of Agricultural Technology." Chinese Economic Studies, 13:3 (Spring 1980): 20-45.

1242. Reeder, J.A. "A Small Study of a Big Market in the People's Republic of China--The Free Market System." The Columbia Journal of World Business, 18 (Winter 1983): 74-80.

1243. Bonavia, D. "Disappointing Year Due to Weather--Or Policies?" The Far Eastern Economic Review, 109 (Sept. 26, 1980): 59-60.

1244. Chinn, D.L. "A Calorie-Arbitrage Model of Chinese Grain Trade." Journal of Developmental Studies, 17:4 (July 1981): 357-70.

1245. Putterman, L. "Extrinsic Versus Intrinsic Problems of Agricultural Cooperation: Anti-Incentivism in Tanzania and China." The Journal of Development Studies, 21 (Jan. 1985): 176-204.

1246. "More on China's New Family Contract System." Monthly Review, 35 (April 1984): 43.

Agricultural Supply and Demand
[711]

1247. Rada, E.L. "Food Policy in China: Recent Efforts to Balance Supplies and Consumption Requirements." Asian Survey, 23 (April 1983): 518-35.

1248. "Counties Double Output Value." Beijing Review, 27 (Jan. 30, 1984): 10-11.

1249. Lu, L. and Liu, Z. "Food Crops Providing Stable Staples." Beijing Review, 29 (Jan. 20, 1986): 16-18.

1250. "Why China's Food Shortages Won't Go Away." Business Week (June 15, 1981): 60.

1251. Skinner, G.W. "Vegetable Supply and Marketing in Chinese Cities." The China Quarterly, 76 (Dec. 1978): 733-93.

1252. Walker, K.R. "Interpreting Chinese Grain Consumption Statistics." The China Quarterly, 92 (Dec. 1982): 576-88.

1253. Delfs, R. "Economic Monitor: Rural Output Rises Further." The Far Eastern Economic Review, 131 (Jan. 16, 1986): 104-5.

1254. Chinn, D.L. "Calorie-Arbitrage Model of Chinese Grain Trade." Journal of Developmental Studies, 17:4 (July 1981): 357-70.

1255. Kueh, Y.Y. "China's Foodgrain Production, Consumption, and Trade: Recent Trends & Prospects." Rivista Internazionale di Scienze Economiche e Commerciali, 31:9 (Sept. 1984): 910-26.

Agricultural Situation and Outlook
[712]

1256. Surls, F.M. "New Directions in China's Agricultural Imports." American Journal of Agricultural Economics, 62 (May 1980): 349-55.

1257. Hsu, R.C. "Agricultural Mechanization in China: Policies, Problems, and Prospects." Asian Survey, 19 (May 1979): 436-49.

1258. "Spring on the Grassland." [cattle industry] Beijing Review, 24 (June 15, 1981): 22-28.

1259. "Rich Summer Harvest." Beijing Review, 24 (Aug. 31, 1981): 5-6.

1260. "Good Prospects for Agriculture." Beijing Review, 25 (Feb. 8, 1982): 5-7.

1261. "Change in Peasants' Mentality." Beijing Review, 26 (Jan. 10, 1983): 6-7.

1262. Wang, D. "Will Peasants Be Polarized by Changes?" Beijing Review, 28 (Nov. 18, 1985): 4-5.

1263. "For China, Another Great Leap Nowhere." Business Week (May 23, 1983): 77.

1264. Timmer, C.P. "China and the World Grain Market." Challenge, 24:4 (Sept./Oct. 1981): 13-21.

1265. Myers, R.H. "Wheat in China--Past, Present and Future." The China Quarterly, 74 (June 1978): 297-333.

1266. ———. Reply with rejoinder by E.B. Vermeer. The China Quarterly, 81 (March 1980): 137-41.

1267. Walker, K.R. "Chinese Agriculture During the Period of the Readjustment, 1978-83; Statistics." The China Quarterly, 100 (Dec. 1984): 783-812.

1268. Stone, B. "The Basis for Chinese Agricultural Growth in the 1980s and 1990s: A Comment on Document No. 1, 1984." The China Quarterly, 101 (March 1985): 114-21.

1269. Li Erhuang. "Can't Suzhou Prefecture's Two-Three System Increase Yield?" Chinese Economic Studies, 15:2 (Winter 1981/82): 93-98.

1270. Maxwell, N. "China Model for Agriculture in the 1980s." Development, 22:2-3 (1980): 23-30.

1271. "Against the Grain." The Economist, 275 (April 12, 1980): 58.

1272. "Are the Good Times Ending?" The Economist, 289 (Dec. 24, 1983): 39-40.

1273. "The Answer Lies in the Soil." The Economist, 293 (Oct. 13, 1984): 78+.

1274. Bonavia, D. "At the Root of the Problem." The Far Eastern Economic Review, 104 (May 11, 1979): 49-50.

1275. Delfs, R. "Critical Impact of China's Grain Card." The Far Eastern Economic Review, 122 (Oct. 6, 1983): 56-58.

1276. Delfs, R. "Agricultural Yields Rise, But the Boom Cannot Last." The Far Eastern Economic Review, 126 (Dec. 13, 1984): 66-68.

1277. Bonavia, D. "Reaping Only What Is Sown." The Far Eastern Economic Review, 131 (Jan. 2, 1986): 28-29.

1278. Pannell, C.W. "Recent Chinese Agriculture." The Geographic Review, 75 (April 1985): 170-85.

1279. "China's Agricultural Revolution." OECD Observer (May 1985): 32.

1280. Stavis, B. "Some Initial Results of China's New Agricultural Policies." World Development, 13:12 (Dec. 1985): 1299-1305.

Agricultural Policy,
Domestic and International
[713]

1281. Tang, A.M. "China as a Factor in the World Food Situation." American Journal of Agricultural Economics, 64:2 (May 1982): 323-31.

1282. Johnson, D.G. "Agriculture in the Centrally Planned Economies." American Journal of Agricultural Economics, 64 (Dec. 1982): 845-53.

1283. Hsu, R. "Grain Procurement and Distribution in China's Rural Areas; Post-Mao Policies and Problems." Asian Survey, 24 (Dec. 1984): 1229-46.

1284. "New Policy for Building Northwest China." Beijing Review, 24 (Jan. 5, 1981): 6-7.

1285. "Chinese Leaders Put Agriculture on the Agenda." Beijing Review, 24 (Aug. 24, 1981): 18-20.

1286. "Strategy for Developing Agriculture." Beijing Review, 25 (Feb. 15, 1982): 26-28.

1287. Zhang Jinfu. "Upholding Planned Economy in Agriculture." Beijing Review, 25 (March 22, 1982): 18-22.

1288. "Program for Current Agricultural Work." Beijing Review, 25 (June 14, 1982): 21-27.

1289. "Chinese-Type Modernization: The Way for Agriculture." Beijing Review, 26 (Jan. 24, 1983): 14-17.

1290. "Becoming Well-Off with Information." Beijing Review, 27 (Dec. 10, 1984): 9-10.

1291. Zhao Ziyang. "Why Relax Agricultural Price Controls?" Beijing Review, 28 (Feb. 18, 1985): 16-18+.

1292. "U.S.-China Trade: A Decade of Development and Prospects for Growth." Business America, 5 (June 28, 1982): 2+.

1293. Nolan, P. "De-Collectivism of Agriculture in China, 1979-82: A Long-Term Perspective." Cambridge Journal of Economics, 7 (Sept./Dec. 1983): 381-403.

1294. Fan, C. and Fan, L.-S. "Some Recent Developments in Chinese Incentive Schemes in Agriculture." Canadian Journal of Agricultural Economics, 28:2 (July 1980): 83-86.

1295. Reynolds, B.L. "Two Models of Agricultural Development: A Context for Current Chinese Policy." The China Quarterly, 76 (Dec. 1978): 842-72.

1296. Friedman, E. "Politics of Local Models, Social Transformation and State Power Struggles in the People's Republic of China: Tachai and Teng Hsiao-P'ing." The China Quarterly, 76 (Dec. 1978): 873-90.

1297. Travers, L. "Post-1978 Rural Economic Policy and Peasant Income in China." The China Quarterly, 98 (June 1984): 241-59.

1298. Surls, F.M. "Agricultural Policy and Growth." [review article] The China Quarterly, 100 (Dec. 1984): 866-69.

1299. Yung-kuei, C. "Treat Farmland Capital Construction as a Great Socialist Task." Chinese Economic Studies, 12:1-2 (Fall/Winter 1978/79): 21-36.

1300. Ch'iu-li, Y. "Mobilize the Whole Party: Fight a Decisive Battle for Three Years: Strive Hard to Basically Realize Agricultural Modernization." Chinese Economic Studies, 12:1 (Fall/Winter 1978/79): 50-82.

1301. He Jianzhang and Wu Kaitai. "A Few Problems Concerning How to Accelerate the Development of Our Country's Agriculture." Chinese Economic Studies, 13:1-2 (Fall/Winter 1979/80): 86-104.

1302. Stavis, B. and Meisner, M. "China's Cropping System Debate: Introduction." Chinese Economic Studies, 15:2 (Winter 1981/82): 4-36.

1303. Mu Jiajun and Ji Jincheng. "Three Times Three Equals Nine Is Not as Good as Two Times Five Equals Ten: An Investigation of Songjiang Country's Experiment to Convert Triple Cropping into Double Cropping." Chinese Economic Studies, 15:2 (Winter 1981/82): 37-44.

1304. Xiong Yi. "Viewpoints & Suggestions on the Southern Jiangsu Cropping System." Chinese Economic Studies, 15:2 (Winter 1981/82): 45-57.

1305. Lin Jiyang and Liu Peiti. "We Must Reform the Cropping System to Suit Local Conditions." Chinese Economic Studies, 15:2 (Winter 1981/82): 58-63.

1306. Huang Pinfu. "Further Discussions on Viewpoints and Suggestions on the Southern Jiangsu Cropping System." Chinese Economic Studies, 15:2 (Winter 1981/82): 64-71.

1307. Zhou Zhengdu. "A View of the Two-Three Cropping System in Suzhou Prefecture." Chinese Economic Studies, 15:2 (Winter 1981/82): 72-76.

1308. Liu Sunhao and Wang Zaide. "Decisions Should Be Taken on the Mainstream Reforms in Cropping Systems." Chinese Economic Studies, 15:2 (Winter 1981/82): 82-89.

1309. Wang Yongnian. "The Two-Three System Made Us Peasants Lose Out." Chinese Economic Studies, 15:2 (Winter 1981/82): 99-101.

1310. Yang Ruichun and Zheng Lizhi. "Looking at the Two-Three System from the Points of View of a Survey of Agricultural Production Costs and of the Records of Food Grain Distribution." Chinese Economic Studies, 15:2 (Winter 1981/82): 107-12.

1311. Zhang Liufang. "We Must Adjust and Reform the Cropping System When the Gains Cannot Offset the Losses." Chinese Economic Studies, 15:2 (Winter 1981/82): 113-16.

1312. Zang Danan; Sun Lingen; and Sun Yousheng. "Does Criticizing the Two-Three System Amount to Ignoring the Broader Picture?" Chinese Economic Studies, 15:2 (Winter 1981/82): 113-16.

1313. Lu Shijian. "In Reforming the Cropping System, We Must Seek Truth from Facts." Chinese Economic Studies, 15:2 (Winter 1981/82): 115-32.

1314. Zhang, Z. and Song, D. "Separation of Government Administration from Commune Management Is a Need of

Rural Economic Development and of Building Political Power." Chinese Economic Studies, 17:3 (Spring 1984): 76-83.

1315. "Can Deng Get China to Work for Itself?" The Economist, 293 (Oct. 27, 1984): 75.

1316. Liu, M. "Element of Risk." The Far Eastern Economic Review, 105 (July 6, 1979): 36-38.

1317. Liu, M. "Firming Up the Foundations." The Far Eastern Economic Review, 105 (July 13, 1979): 45-47.

1318. Bonavia, D. "Remodeling the Cooperative Society." The Far Eastern Economic Review, 120 (April 28, 1983): 52-53.

1319. do Rosario, L. "No More the Leader." The Far Eastern Economic Review, 131 (March 6, 1986): 32-33.

1320. Sicular, T. "Agricultural Planning in China: The Case of Lee Willow Team No. 4." Food Research Institute Studies, 20:1 (1986): 1-24.

1321. "Institution Reforms [Editorial: production responsibilities for farmlands in China]." Journal of Commerce and Commercial, 357 (July 25, 1983): 4A.

1322. Kueh, Y.-Y. "China's New Agricultural-Policy-Program: Major Economic Consequences 1979-1983." Journal of Comparative Economics, 8:4 (Dec. 1984): 353-75.

1323. Putterman, L. "Extrinsic vs. Intrinsic Problems of Agricultural Corporation." Journal of Development Studies, 21:2 (Jan. 1985): 175-204.

1324. Blecher, H. "The Structure and Contradictions of Productive Relations in Socialist Agrarian 'Reform': A Framework for Analysis and the Chinese Case." Journal of Development Studies, 22 (Oct. 1985): 104-26.

1325. Friedman, E. "Beijing and the Countryside." [review article] Problems of Communism, 28 (Sept.-Dec. 1979): 77-80.

1326. Etienne, G. "Rural Development in China and Its Impact on Foreign Trade." Rivista Internazionale di Scienze Economiche e Commerciali, 28:9 (Sept. 1981): 831-51.

1327. "Mao's Rural Strategies: What Went Wrong?" Science and Society, 49:1 (Spring 1985): 101-7.

1328. Selden, M. "Logic--and Limits--of Chinese Socialist Development." World Development, 11 (Aug. 1983): 631-37.

1329. Stavis, B. "Some Initial Results of China's New Agricultural Policies." World Development, 13 (Dec. 1985): 1299-1305.

Agricultural Finance
[714]

1330. "Using Foreign Investment in Agriculture." Beijing Review, 27 (Feb. 20, 1984): 18-20.

1331. "Foreign Capital Boosts Agriculture." Beijing Review, 28 (Jan. 7, 1985): 40-41.

1332. "Foreign Funds Aid Farm Production." Beijing Review, 28 (Nov. 18, 1985): 29-30.

Agricultural Marketing and Agribusiness
[715]

1333. "Different Prices for Farm Products." Beijing Review, 26 (Aug. 29, 1983): 21-24.

1334. Li Yongzhen. "Peasants' Enthusiasm for Science." Beijing Review, 27 (March 12, 1984): 24-31.

1335. "Peasants Turn to Science for Help." Beijing Review, 28 (June 10, 1985): 22-25.

1336. "Farmers Take on New Business." Beijing Review, 28 (Feb. 11, 1985): 8-9.

1337. "Legal Firms Spur Rural Economy." Beijing Review, 28 (Nov. 4, 1985): 9-10.

1338. "Agro-Science Enters Golden Age." Beijing Review, 29 (Jan. 27, 1986): 6-7.

1339. Skinner, G.W. "Rural Marketing in China: Repression and Revival." The China Quarterly, 103 (Sept. 1985): 393-413.

1340. Lu, B. and Yuan, Z. "On Several Problems in the Current Situation with Regard to the Requisition and Procurement of Agricultural and Sideline Products." Chinese Economic Studies, 18:3 (Spring 1985): 3-19.

1341. "Mark-Ups in the Market Place." The Economist, 273 (Nov. 10, 1979): 76+.

1342. "Grab a Quick Yuan While You Can." The Economist, 285 (Oct. 30, 1982): 45-46.

1343. "The Farming Base." The Economist, 297 (Dec. 21, 1985): 66-67.

1344. Delfs, R. "Back to the Market." The Far Eastern Economic Review, 128 (May 30, 1985): 77-79.

Agricultural Reforms,
Responsibility System
[716]

1345. Wiens, T.B. "Price Adjustment, the Responsibility System, and Agricultural Productivity." American Economic Review, 73 (May 1983): 319-24.

1346. Pang, C.M. and De Boer, A.J. "Management Decentralization on China's State Farms." American Journal of Agricultural Economics, 65 (Nov. 1983): 657-74.

1347. Leeming, F. "Progress Towards Triple-Cropping in China." Asian Survey, 19 (May 1979): 450-67.

1348. Zweig, D. "Opposition to Change in Rural China: The System of Responsibility and People's Communes." Asian Survey, 23 (July 1983): 879-900.

1349. "System of Responsibility in Agricultural Production." Beijing Review, 24 (March 16, 1981): 3-4.

1350. "System That Mobilizes Peasant's Enthusiasm." Beijing Review, 24 (April 27, 1981): 6-8.

Agriculture

1351. "Popularization of Agrotechniques." Beijing Review, 24 (May 11, 1981): 8-9.

1352. "Now the Peasants Can Decide." Beijing Review, 24 (Aug. 31, 1981): 15-18.

1353. "Scientific Farming Encouraged." Beijing Review, 25 (March 8, 1982): 7-8.

1354. "Income Gap." Beijing Review, 25 (June 21, 1982): 3-4.

1355. "Putting Rural Surplus Workers to Work." Beijing Review, 26 (Dec. 19, 1983): 6-7.

1356. "Rural Responsibility System in Theory and Practice." Beijing Review, 27 (April 9, 1984): 27-29.

1357. Lu Yun. "Rural Responsibility System." Beijing Review, 27 (Oct. 29, 1984): 18+; (Nov. 5, 1984): 23-25; (Nov. 12, 1984): 24-27; (Nov. 19, 1984): 20-22; (Dec. 3, 1984): 23-27; (Dec. 10, 1984): 18-21.

1358. Lu Yun. "Rural Responsibility System: Gap Between Rich and Poor Is Bridged." Beijing Review, 27 (Nov. 12, 1984): 24-27.

1359. Lu Yun. "Rural Responsibility System: Will Farm Mechanization Be Slowed?" Beijing Review, 27 (Nov. 19, 1984): 20-22.

1360. "China: Gaining Efficiency Through Western Ways." Business Week (May 24, 1982): 172.

1361. "Returning Incentives to the Farmer." China Business Review, 9 (Nov./Dec. 1982): 16-17.

1362. "Rivers of Waste." China Business Review, 10 (July/Aug. 1983): 18-20.

1363. "Incentive Farming." China Business Review, 10 (Nov./Dec. 1983): 12-14.

1364. Kallgren, J. "The Concept of Decentralization in Document No. 1, 1984." The China Quarterly, 101 (March 1985): 104-8.

1365. Ma Chuandong and Tian Jiasen. "Implement the 'Four Unifications' and Bring into Full Play the Role of

Farm Machinery." *Chinese Economic Studies*, 13:1-2 (Fall/Winter 1979/80): 22-39.

1366. Shi Qian, et al. "Investigative Report: A Few Problems Involved in the Effort to Speed Up Agricultural Mechanization as Seen from the Experience at Wuming." *Chinese Economic Studies*, 13:1-2 (Fall/Winter 1979/80): 58-73.

1367. Tang, A.M. "China's Agricultural Legacy." *Economic Development and Cultural Change*, 28 (Oct. 1979): 1-22.

1368. Bonavia, D. "China Rediscovers the Family Farm." *The Far Eastern Economic Review*, 112 (June 19-25, 1981): 56-57.

1369. O'Leary, G. and Watson, A. "Back to the Family in the Field." *The Far Eastern Economic Review*, 115 (Feb. 26-March 4, 1982): 84+.

1370. Delfs, R. "It's Getting Much More Efficient Down on the Farm." *The Far Eastern Economic Review*, 118 (Oct. 1-7, 1982): 49-50.

1371. Johnson, G. "Responsibility Reaps Rewards." *The Far Eastern Economic Review*, 122 (Oct. 6, 1983): 55-57.

1372. Chinn, D.L. "Team Cohesion and Collective Labor Supply in Chinese Agriculture." *Journal of Comparative Economics*, 3:4 (Dec. 1979): 375-94.

1373. Cremer, J. "On the Efficiency of a Chinese-Type Work-Point System." *Journal of Comparative Economics*, 6:4 (Dec. 1982): 343-52.

1374. Chinn, D.L. "Diligence and Laziness in Chinese Agricultural Production Teams." *Journal of Developmental Economics*, 7:3 (Sept. 1980): 331-44.

1375. Lardy, N.R. "Agricultural Reforms in China." *Journal of International Affairs*, 39 (Winter 1986): 91-104.

1376. Hinton, W.H. "Trip to Fengyang County: Investigating China's New Family Contract System." *Monthly Review*, 35 (Nov. 1983): 1-28.

1377. Woodward, D. "New Direction for China's State Farms." Pacific Affairs, 55 (Summer 1982): 231-51.

1378. Watson, A. "New Structures in the Organization of Chinese Agriculture: A Variable Model." Pacific Affairs, 57 (Winter 1984/85): 621-45.

1379. Crook, F.W. "Motivating China's Farmers." [review article] Problems of Communism, 32 (Sept./Oct. 1983): 64-71.

1380. Chenn, D.L. "Cooperative Farming in North China." Quarterly Journal of Economics, 94:2 (March 1980): 279-97.

1381. "Communes Show Chinese Hard at Work." Travel Weekly, 43 (Feb. 13, 1984): 19.

1382. "End of an Era for China's Farm Communes." U.S. News & World Report, 94 (Jan. 17, 1983): 30.

1383. Watson, A. "Agriculture Looks for Shoes That Fit: The Productions Responsibility System and Its Implications." World Development, 11 (Aug. 1983): 705-30.

Rural Reforms
[717]

1384. Du, R. "Second Stage Rural Structural Reform." Beijing Review, 28 (June 24, 1985): 15-17+.

1385. "Economic Growth and Rural Development." Beijing Review, 29 (March 10, 1986): 14-21.

1386. Kueh, Y.Y. "The Economics of the 'Second Land Reform' in China." The China Quarterly, 101 (March 1985): 122-31.

1387. Perry, E.J. "Rural Violence in Socialist China." China Quarterly, 103 (Sept. 1985): 414-40.

1388. Watson, A. "New Structures in the Organization of Chinese Agriculture: A Variable Model." Pacific Affairs, 57 (Winter 1984/85): 621-45.

1389. Burns, J.P. "Local Cadre Accommodation to the 'Responsibility System' in Rural China." Pacific Affairs, 58 (Winter 1985/86): 607-25.

1390. Fewsmith, J. "Rural Reform in China: Stage Two." Problems of Communism, 34 (July/Aug. 1985): 48-55.

Rural Situation
[718]

1391. "More Income for Peasants; Changes in Poor Villages." Beijing Review, 24 (Jan. 19, 1981): 5-6.

1392. "Rural Policy: Special Feature." Beijing Review, 24 (Jan. 19, 1981): 19-29.

1393. "Diversified Rural Economy." Beijing Review, 24 (May 4, 1981): 5-6.

1394. Du Runsheng. "Good Beginning for Reform of Rural Economic System." Beijing Review, 24 (Nov. 30, 1981): 15-20.

1395. Tian Yun. "Life and Work for Youth in Fujian Countryside." Beijing Review, 25 (March 15, 1982): 18-26.

1396. "Rapid Growth of Small Enterprises in Rural Areas." Beijing Review, 25 (Sept. 13, 1982): 7-8.

1397. "Strengthen Rural Technical Work." Beijing Review, 26 (March 7, 1983): 10-11.

1398. "Rural Women and the New Economic Policies." Beijing Review, 26 (March 7, 1983): 18-20.

1399. "New Achievements in Rural Economy." Beijing Review, 26 (Sept. 5, 1983): 6-7.

1400. "Assisting the Rural Poor." Beijing Review, 26 (Sept. 19, 1983): 23-27.

1401. "Sample Survey of Peasant Household Incomes and Expenditures." Beijing Review, 26 (Oct. 24, 1983): 22-25.

1402. "New Productive Forces in Rural Areas." Beijing Review, 27 (Feb. 27, 1984): 4-5.

1403. "Developing Rural Commodity Production." Beijing Review, 27 (Feb. 27, 1984): 16-21+.

Agriculture

1404. "Booming Rural Science Associations." Beijing Review, 27 (March 26, 1984): 11.

1405. "Chinese Peasants Favor Small Towns." Beijing Review, 28 (April 1, 1984): 4-5.

1406. "Rural Businesses Providing Jobs." Beijing Review, 27 (April 16, 1984): 7-8.

1407. "Explaining China's Rural Economic Policy." Beijing Review, 27 (April 30, 1984): 16-21.

1408. Du Runsheng. "China's Countryside under Reform." Beijing Review, 27 (Aug. 13, 1984): 16-21.

1409. "Rural Women Gain Economic Status." Beijing Review, 27 (Oct. 15, 1984); 9-10.

1410. "Circular Outlines Plan to Help Poor." Beijing Review, 27 (Nov. 5, 1984): 7-8.

1411. "Peasants Become Rural Workers." Beijing Review, 27 (Nov. 12, 1984): 7-8.

1412. Lu Yun. "Rural Responsibility System: Rural Township Enterprises Flourish." Beijing Review, 27 (Dec. 10, 1984): 18-21.

1413. "Nuclear Families Dominate Countryside." Beijing Review, 28 (Jan. 7, 1985): 42-43.

1414. "Government Aids Poor Farmers." Beijing Review, 18 (Jan. 28, 1985): 9-10.

1415. "Further Explorations of Small Towns." Beijing Review, 28 (April 8, 1985): 24-26; (April 29, 1985): 22-23; (May 27, 1985): 20-22.

1416. "Youth Encouraged to Become Pioneers." Beijing Review, 28 (May 13, 1985): 6-7.

1417. "Rural Young People Change Their Outlook." Beijing Review, 28 (May 13, 1985): 23-25.

1418. Fei Xiaotong. "Surplus Rural Labour Put to Work." Beijing Review, 28 (May 27, 1985): 20-22.

1419. Saith, A. "Economic Incentives for the One-Child Family in Rural China." The China Quarterly, 87 (Sept. 1981): 492-500.

1420. Vermeer, E.B. "Income Differentials in Rural China." The China Quarterly, 89 (March 1982): 1-33; 92 (Dec. 1982): 706-13.

1421. "Circular of the Central Committee of the Chinese Communist Party on Rural Work During 1984." The China Quarterly, 101 (March 1985): 132-42.

1422. Zweig, D. "Prosperity and Conflict in Post-Mao Rural China." The China Quarterly, 105 (March 1986): 1-18.

1423. Feng, J. "Bring Out the Superiority of the System of Contracted Responsibilities on the Household Basis." Chinese Economic Studies, 17:3 (Spring 1984): 18-26.

1424. Du, R. "New Developments in the Contracting System of United Production and the Cooperative Economy in the Countryside." Chinese Economic Studies, 17:4 (Summer 1984): 16-39.

1425. Blecher, M. and Meisner, M. "Economic Growth and Equality in Rural China." Comparative Political Studies, 13 (Jan. 1981): 505-26.

1426. Bonavia, D. "Peking Seeks New Bricks to Fill a Hole in the Wall." The Far Eastern Economic Review, 103 (Jan. 26, 1979): 34-35.

1427. Bonavia, D. "Reality of Reform." The Far Eastern Economic Review, 114 (Oct. 9-15, 1981): 75-76.

1428. Bonavia, D. "Facing Up to the Classic LDC Dilemma." The Far Eastern Economic Review, 127 (March 21, 1984): 82+.

1429. Delfs, R. "Breaking Out of the Mould into a Golden Ricebowl." The Far Eastern Economic Review, 126 (Dec. 13, 1984): 70-71.

1430. Delfs, R. "The Rural Uprising." The Far Eastern Economic Review, 129 (July 11, 1985): 56-57.

1431. Delfs, R. "The Delta Factor." [rural industries] The Far Eastern Economic Review, 129 (July 18, 1985): 93-95.

1432. Tan, K.C. "Revitalized Small Towns in China." Geographical Review, 76 (April 1986): 138-48.

1433. "Chinese Communes Offer More to Workers." Journal of Commerce, 353 (Aug. 19, 1982): 3A.

1434. Griffin, K. and Saith, A. "Pattern of Income Inequality in Rural China." Oxford Economic Papers, 34 (March 1982): 172-206.

1435. Chan, A. and Unger, J. "Grey and Black: The Hidden Economy of Rural China." Pacific Affairs, 55 (Fall 1982): 452-71.

1436. O'Leary, G. and Watson, A. "Role of the People's Commune in Rural Development in China." Pacific Affairs, 55 (Winter 1982/83): 593-612.

1437. Griffin, K. and Griffin, K. "Institutional Change and Income Distribution in the Chinese Countryside." Oxford Bulletin of Economics and Statistics, 45:3 (Aug. 1983): 223-48.

1438. "Despite Rural China's Gains, Poverty Grips Some Regions." New York Times, 134 (Dec. 18, 1984): A1.

1439. Riskin, C. "Intermediate Technology in China's Rural Industries." World Development, 6:11-12 (Nov./Dec. 1978): 1297-1311.

1440. Maxwell, N. "A 'Paupers' Co-Op' Twenty-Five Years On: Capital Formation in Rural China." World Development, 7:4-5 (April/May 1979): 433-46.

1441. Blecher, M. "Peasant Labour for Urban Industry: Temporary Contract Labour, Urban-Rural Balance and Class Relations in a Chinese County." World Development, 11:8 (Aug. 1983): 731-45.

1442. Blecher, M. "Inequality and Socialism in Rural China: A Conceptual Note." World Development, 13 (Jan. 1985): 115-21.

Rural Industry
[719]

1443. "Rural Enterprises Take on Economy." Beijing Review, 28 (Dec. 16, 1985): 8-9.

1444. Xue, M. "Rural Industry Advances Amidst Problems." Beijing Review, 28 (Dec. 16, 1985): 18-21.

NATURAL RESOURCES
[720]

Natural Resources, Oil,
Mining, Metals
[721]

1445. "Status of Metals Confused as China Embarks on Trade Recentralization." American Metal Market, 92 (March 28, 1984): 1.

1446. "Steady Rise in U.S. Oilfield Equipment to China Seen: 1984 Sales to Hit $100 M." American Metal Market, 92 (Aug. 20, 1984): 4.

1447. "China Government Expands Rare Earth Industry." American Metal Market, 93 (Sept. 13, 1985): 1.

1448. "More Shanxi Coal for Export." Beijing Review, 24 (March 2, 1981): 6-7.

1449. "Prospects of China's Coal Industry." Beijing Review, 26 (Sept. 12, 1983): 14-16.

1450. "Chinese-Foreign Oil Co-Operation." Beijing Review, 26 (Dec. 19, 1983): 5-6.

1451. "Geothermal Energy Lights Up Tibet." Beijing Review, 27 (Aug. 27, 1984): 28-29.

1452. Shi Yan. "China's Burgeoning Oil Industry." Beijing Review, 27 (Sept. 10, 1984): 17-20.

1453. "Bright Prospects for Offshore Oil." Beijing Review, 28 (Sept. 23, 1985): 29-30.

Natural Resources

1454. "Scrambling for China's Oil." Business Week (May 31, 1982): 94-95.

1455. "Arco's Deal with China Is a Tough Act to Follow." Business Week (Oct. 4, 1982): 43-44.

1456. "Behind BP's Oil Deal with Beijing." Business Week (May 30, 1983): 31-32.

1457. "Oxy May Have a Long Wait Before Profits Surface in China." Business Week (Aug. 22, 1983): 32-33.

1458. "The X-Factor: Is Chinese Oil Worth China's Price?" Business Week (March 19, 1984): 94+.

1459. "Geology of China's Oil." The Economist, 270 (March 3, 1979): 100-2.

1460. "Chinese Oil--Offshore Checkers." The Economist, 280 (Sept. 12, 1981): 70-71.

1461. "China's Offshore Oil Will Fuel Hong Kong's Stock Market." The Economist, 283 (April 24-30, 1982): 105-6.

1462. "Help Us Dig It Out." The Economist, 287 (April 23-29, 1983): 73.

1463. Scott, W.E. "China Seeks Help of International Oil Firms." Energy International, 17 (June 1980): 35-38.

1464. "China: An Economic Evaluation of Ore Reserve Losses at Underground Mines." Engineering & Mining Journal, 186 (May 1985): 40.

1465. "China Oil Rush: First Lease for Exploration on Land Awarded to Aussies." Engineering News-Record, 214 (June 6, 1985): 15.

1466. Munthe-Kaas, H. and Lasuriat, G. "China Learns from Norway." The Far Eastern Economic Review, 103 (Feb. 23, 1979): 96-97.

1467. Lauriat, G. and Liu, M. "Pouring Trouble on Oily Waters." The Far Eastern Economic Review, 105 (Sept. 28, 1979): 18-21.

1468. Chanda, N. "China Calls in the Foreign Rigs." The Far Eastern Economic Review, 105 (Sept. 28, 1979): 21.

1469. Lauriat, G. "Another Coming Conflict of Comrades Ahead." The Far Eastern Economic Review, 106 (Oct. 5, 1979): 58-59.

1470. Rowley, A. "Cashing In on China's Oil Boom." The Far Eastern Economic Review, 110 (Nov. 14-20, 1980): 52-54.

1471. Loong, P. "Oil Eases China's Deficit." The Far Eastern Economic Review, 111 (Jan. 30-Feb. 5, 1981): 44-45.

1472. Lee, M. "China Rigs the Oil Game." The Far Eastern Economic Review, 112 (April 3-9, 1981): 32-33.

1473. Rivers, C. "China's 10-Billion-Tonne Offshore Oil Bonanza." The Far Eastern Economic Review, 114 (Oct. 2-8, 1981): 57-58.

1474. Delfs, R. "Getting Together." [petroleum industry] The Far Eastern Economic Review, 114 (Dec. 4-10, 1981): 86-88.

1475. Ignatius, A. "Waiting for the Offshore Go-Ahead." The Far Eastern Economic Review, 118 (Oct. 1-7, 1982): 63-66.

1476. Ma, T. "China's Watering Holes." The Far Eastern Economic Review, 119 (March 31, 1983): 76-77.

1477. Ma, T. "Way into China Oil." The Far Eastern Economic Review, 120 (June 23, 1983): 86-87.

1478. Ma, T. "Foreigners Too Would Like Some Offshoot Business." The Far Eastern Economic Review, 121 (Aug. 25, 1983): 61-63.

1479. Ma, T. "Target: 200 Million Tonnes by 2000." The Far Eastern Economic Review, 121 (Aug. 25, 1983): 64-65.

1480. Ma, T. "Eighteen of 25 Bidders Are Signed Up Offshore." The Far Eastern Economic Review, 122 (Oct. 6, 1983): 69-71.

1481. Langston, N. "Wells of Uncertainty." The Far Eastern Economic Review, 124 (June 28, 1984): 50-52.

1482. Langston, N. "Problems in the Pipeline; Coal Industry." The Far Eastern Economic Review, 125 (Aug. 16, 1984): 50-51.

1483. Kulkarni, V.G. "Problems of Plenty." [oil] The Far Eastern Economic Review, 126 (Nov. 8, 1984): 98.

1484. Langston, N. and Lee, M. "Elephants' Graveyard--Oil in the South China Sea." The Far Eastern Economic Review, 126 (Dec. 6, 1984): 63-64.

1485. Langston, N. "Round Two, Out of the Corner: The Oil Majors Prepare to Bid Again in China." The Far Eastern Economic Review, 129 (Aug. 1, 1985): 50-51.

1486. Phillips, D.R. "Oil in Chinese Waters." The Geographical Magazine, 56 (Sept. 1984): 444-45.

1487. "Chinese Start Oil-Equipment Buying Spree." Journal of Commerce and Commercial, 364 (May 16, 1985): 3B.

1488. Herschede, F. and Kadhim, M. "China's Petroleum Production and Reserves: Domestic and International Significance." Journal of Energy Development, 5:1 (Autumn 1979): 57-71.

1489. Edmond, M. "Chinese Gasoline Not All That Cheap." National Petroleum News, 74 (June 1982): 19-20.

1490. Edmond, M. "More Chinese Gas to Come? Arco-China Connection the Key." National Petroleum News, 76 (Feb. 1984): 28.

1491. Lomax, D.F. "The Investment Implications of China's Offshore Oil Development." National Westminster Bank Quarterly Review (Feb. 1986): 50-69.

1492. Emerson, J.D. "People's Republic of China Offshore Reserves Put at 39 Billion Bbls." Ocean Industry, 17 (March 1982): 32+.

1493. "China Pushing Expansion of Oil and Gas." Oil & Gas Journal, 77 (April 23, 1979): 26-28.

1494. "China Trims Oil Production Goals." Oil & Gas Journal, 77 (Sept. 3, 1979): 36-38.

1495. "Successes Spur Chinese Search Onshore and Off." Oil & Gas Journal, 77 (Nov. 26, 1979): 19-21.

1496. "China's Petroleum Surplus May Vanish in the 1980s." Oil & Gas Journal, 78 (Oct. 6, 1980): 27-32.

1497. "S. China Sea Seen Big Oil, Gas Region." Oil & Gas Journal, 78 (Nov. 3, 1980): 44-46.

1498. "Lagging Production Trims China's Oil Exports." Oil & Gas Journal, 79 (March 23, 1981): 58-59.

1499. "China's Oil Outlook Grows Brighter." Oil & Gas Journal, 27 (May 25, 1981): 120-22.

1500. "Foreign Assistance Paying Off." Oil & Gas Journal, 79 (Nov. 9, 1981): 124-25.

1501. Shanweng, W., et al. "Habitat of Oil and Gas Fields in China." Oil & Gas Journal, 80 (June 14, 1982): 119-21+.

1502. Enright, R.J. "Chinese Shelf May Generate Biggest Offshore Flurry Since North Sea." Oil & Gas Journal, 80 (Dec. 13, 1982): 55-60.

1503. Zuoxiang, A. "China Develops Natural Gas Industry." Oil & Gas Journal, 80 (Sept. 6, 1982): 145-46.

1504. "China: Special Report." Oil & Gas Journal, 80 (Dec. 13, 1982): 55-91+.

1505. Zhifang, L. "Water Injection in China: Waterflooding the Oil Reservoirs." Oil & Gas Journal, 80 (Dec. 13, 1982): 61-62+.

1506. Zaiyi, T., et al. "Sedimentary Facies, Oil Generation in Meso-Cenozoic Continental Basins in China." Oil & Gas Journal, 81 (May 16, 1983): 120-26.

1507. Parker, M.A. "Far East Action Seen Tied to Oil Price, China, and Market." Oil & Gas Journal, 81 (May 16, 1983): 128-29.

Natural Resources

1508. Renpu, W. "Status and Prospects: Petroleum Production Technology in China." Oil & Gas Journal, 81 (Aug. 29, 1983): 45-50.

1509. "Drilling Activity Off China Predicted to Surge." Oil & Gas Journal, 81 (Oct. 17, 1983): 48+.

1510. "More Tracts Awarded Off China: Action to Build." Oil & Gas Journal, 81 (Dec. 5, 1983): 80-81.

1511. Jones, G.C.L., et al. "The Economics of Marginal Offshore Oil Discoveries in China." Oil & Gas Journal, 82 (March 12, 1984): 79-85.

1512. "Sale of Surplus U.S. Rigs to China Suggested." Oil & Gas Journal, 82 (April 2, 1984): 66-67.

1513. "China's 1990 Oil Output Goal Seen Attainable." Oil & Gas Journal, 83 (Oct. 7, 1985): 50-51.

1514. "Arco Due No Equity Role in China Gas Project." Oil & Gas Journal, 85 (Jan. 14, 1985): 52.

1515. "China Offers More Tracts: ACT Group Tests Strike." Oil & Gas Journal, 83 (Jan. 14, 1985): 55-77.

1516. "Occidental, China Sign Oil Development Accord." Oil Daily (July 2, 1985): 1.

1517. "China's Oil and Energy Industries." Petroleum Economist, 48 (Nov. 1981): 476-80+.

1518. "Foreign Oil Participation." Petroleum Economist, 48 (Nov. 1981): 489-90+.

1519. Segal, J. "Need for More Oil Exploration." Petroleum Economist, 48 (Nov. 1981): 495-98.

1520. "Offshore Exploration." Petroleum Economist, 50 (Nov. 1983): 418-22.

1521. Fridley, D. and Johnson, T.M. "China's Growing Petroleum Trade." Petroleum Economist, 52 (Nov. 1985): 387-92.

1522. "China, Venezuela Study Cooperation Accord [with OPEC]." Platt's Oilgram, 63 (March 27, 1985): 1.

1523. Woodard, K. "China and Offshore Energy." Problems of Communism, 30 (Nov./Dec. 1981): 32-45.

1524. Smock, D. "China Thaw Opens Vast Deposits of Metals, Minerals." Purchasing, 86 (Feb. 21, 1979): 12-13.

1525. Mark, J. "Spotty Offshore Drilling Results Force Chinese to Probe Companies' Interest in Offshore Oil." The Wall Street Journal (Feb. 27, 1985).

1526. "China Set to Sign $500 Million Pact for Gas Field." The Wall Street Journal (Sept. 26, 1985).

1527. "Royal Dutch/Shell Group Quits Venture in China." The Wall Street Journal (Oct. 17, 1985).

1528. Wyllie, R.J.M., et al. "China: A Mineral Giant." World Mining, 32 (July 1979): 11+.

1529. Argall, G.O., Jr. "Mining Operations and Mineral Deposits of China." World Mining, 32 (Oct. 1979): 68-83; 33 (Jan. 1980): 42-43.

1530. Argall, G.O., Jr. "China Encourages Foreign Investment." World Mining, 36 (June 1983): 62-65.

1531. Cheng, L. "Capital Sought for Chinese Development." World Mining Equipment, 8 (March 1984): 48-49.

1532. Scott, R.W. "Oil and Gas in China." World Oil, 189 (Dec. 1979): 55-61.

1533. "First U.S. Platform Rig Working in Chinese Water." World Oil, 190 (Feb. 1, 1980): 76-78.

1534. Yu, M. "Petroleum Exploration in the People's Republic of China." World Oil, 191 (Sept. 1980): 99-103.

1535. Garb, F.A. "Oil and Gas in China." World Oil, 192 (Feb. 1, 1981): 35-41.

1536. Holland, D.S. "Offshore China: High Risk, High Potential." World Oil, 199 (Aug. 1, 1984): 59-60.

Natural Resources 131

Conservation and Pollution
[722]

1537. Smil, V. "Environmental Degradation in China." Asian Survey, 20 (Aug. 1980): 777-88.

1538. Cromley, R.G. "Von Thuenen Model and Environmental Uncertainty." Association of American Geographers. Annals, 72 (Sept. 1982): 404-10.

Energy
[723]

1539. "U.S. Firms Eye China Nuclear Mart." American Metal Market, 92 (May 14, 1984): 1.

1540. "Tackling Rural Energy Shortage." Beijing Review, 24 (Aug. 10, 1981): 6-7.

1541. "Small Hydropower Stations." Beijing Review, 24 (Aug. 10, 1981): 22-29.

1542. "China's Largest Hydropower Project." Beijing Review, 24 (Aug. 31, 1981): 20-27.

1543. Yu Bing. "Energy Development and Conservation." Beijing Review, 25 (March 8, 1982): 20-22.

1544. Shi Wen. "Use of New and Renewable Energy Resources in China." Beijing Review, 25 (April 19, 1982): 18-20.

1545. "Chinese-Style Rural Electrification." Beijing Review, 26 (Feb. 7, 1983): 6.

1546. "How Will China Solve Energy System." Beijing Review, 26 (Aug. 29, 1983): 13-18.

1547. "Priority to Developing Energy." [hydroelectric plants] Beijing Review, 26 (Nov. 7, 1983): 16-19.

1548. "Developing China's Nuclear Power Industry." Beijing Review, 27 (June 18, 1984): 17-20.

1549. "Developing Rural Energy Resources." Beijing Review, 28 (May 27, 1985): 23-25.

1550. Zhou, Z. "The Hongshui River a Mighty Powerhouse."
Beijing Review, 28 (July 1, 1985): 14-18.

1551. Zhou, Z. "Gezhouba Hydroelectric Project Revisited."
Beijing Review, 28 (July 8, 1985): 16-18+.

1552. "China Steps Up Energy Imports." Beijing Review, 29
(April 14, 1986): 30-31.

1553. "Agreement Enhances Prospects for U.S. Firms Bidding
on Hydropower Projects in China." Business America,
5 (Nov. 15, 1982): 33.

1554. "American Edge in China's Hydro Plans." Business Week
(Sept. 17, 1979): 37-38.

1555. "Energy: A Bottleneck in China's Industrial Drive."
Business Week (May 19, 1980): 59+.

1556. "China: Tougher Dealing Over U.S. Nuclear Sales."
Business Week (Aug. 2, 1982): 32-33.

1557. "The U.S. Demands a Fair Shot at Building a Huge Dam."
[Yangzi River] Business Week (May 21, 1984): 54.

1558. "Energy Conservation." China Business Review, 9
(Jan./Feb. 1982): 12-18.

1559. Kambara, T. "China's Energy Development During the
Readjustment and Prospects for the Future." The
China Quarterly, 100 (Dec. 1984): 762-82.

1560. "Reagan Under Pressure to OK Reactor Pact." Christian
Science Monitor, 77 (March 4, 1985): 3.

1561. "China Laid Bare." The Economist, 279 (June 20, 1981):
44-45.

1562. "China's Energy: Grounded." The Economist, 291
(April 7, 1984): 73-74.

1563. "Nuclear Power: Chinese Burn." The Economist, 291
(May 5, 1984): 80+.

1564. "America and China: Dunfission." The Economist, 291
(June 30, 1984): 26-27.

1565. "China's Nuclear Power: Tender Subject." The Economist, 294 (Jan. 12, 1985): 61-62.

1566. "China Gets U.S. Advice on Power." Electrical World, 193 (June 15, 1980): 23-24.

1567. "China Develops Yellow River Hydro Resources." Electrical World, 199 (April 1985): 46-47.

1568. Smil, V. "China Reveals Long-Term Energy Development." Energy International, 15 (Aug. 1978): 23-25+.

1569. Smil, V. "China's Energy Plans Call for Foreign Assistance." Energy International, 15 (Dec. 1978): 27-29.

1570. Smil, V. "China's Perplexing Energy Triangle." Energy International, 16 (June 1979): 25-27.

1571. Djurovic, M. "Mini Hydro Plants Boost China's Power Supply." Energy International, 16 (Nov. 1979): 44-46.

1572. Kharbanda, O.F. "China's Ambitious Development Plans." Energy International, 17 (Feb. 1980): 16-19.

1573. Kharbanda, V.P. and Qureshi, M.A. "Biogas Development in India and the People's Republic of China." Energy Journal, 6:3 (July 1985): 51-65.

1574. Yu, Q. "Cogeneration in the People's Republic of China." Energy Journal, 5 (April 1984): 133-37.

1575. "China Hydro Agreement Gives U.S. Edge for Jobs." Engineering News-Record, 203 (Sept. 6, 1979): 9-10.

1576. "Chinese Move on Huge Water Plan" [Three Gorge Dam] Engineering News-Record, 212 (March 29, 1984): 25-26.

1577. Vinals, J.M. "Energy-Capital Substitution, Wage Flexibility, and Aggregate Output Supply." European Economics Review, 26 (Oct./Nov. 1984): 229-45.

1578. Bonavia, D. "Who Do We Learn From This Time?" [power resources] The Far Eastern Economic Review, 110 (Oct. 10-16, 1980): 64-65.

Agriculture; Natural Resources

1579. Lee, M. "China Light and Power Company, New Vote of Confidence." The Far Eastern Economic Review, 112 (April 17-23, 1981): 48-49.

1580. Delfs, R. "Biggest Producer--The Biggest Consumer." The Far Eastern Economic Review, 114 (Oct. 2-8, 1981): 51-52+.

1581. Delfs, R. "China Spells It Out." The Far Eastern Economic Review, 115 (Jan. 22-28, 1982): 41-42.

1582. Manning, R. "China's Nuclear Boom." The Far Eastern Economic Review, 118 (Nov. 26-Dec. 2, 1982): 19-20.

1583. Delfs, R. "Deeper into the Red." The Far Eastern Economic Review, 120 (April 7, 1983): 62-63.

1584. Delfs, R. "Balance of Power." [French reactors in China] The Far Eastern Economic Review, 120 (May 19, 1983): 80-81.

1585. Nations, R. "Fusion of Ideas." [nuclear cooperation] The Far Eastern Economic Review, 121 (July 28, 1983): 14-15.

1586. Ma, T. "Nuclear Family." [joint venture with Hong Kong] The Far Eastern Economic Review, 122 (Nov. 24, 1983): 62-63+.

1587. Frank, A.D. "Atomic Reaction." [China-U.S. nuclear agreement] Forbes, 133 (June 4, 1984): 153.

1588. Ting Yi-Lan. "Water Is the Chinese Life Force." Geographical Magazine, 52 (Jan. 1980): 257+.

1589. "China Reveals Plans to Export Nuclear Fuel." Journal of Commerce and Commercial, 363 (Jan. 14, 1985): 1A.

1590. "China Debates Energy Options." Journal of Commerce and Commercial, 365 (Sept. 5, 1985): 9A.

1591. "China Signs Three Big Contracts with US Firms." Modern Power Systems, 1 (Feb. 1981): 5.

1592. "Energy-Short Chinese Turn to Hydro, Conservation, Nuclear and Tidal Power." Modern Power Systems, 1 (May 1981): 21-24+.

1593. "China Reaffirms Self-Sufficiency Goal." Oil & Gas Journal, 80 (Jan. 18, 1982): 68-69.

1594. Bin, L. and Chengwu, M. "China Emphasizes Energy Savings in Oil-Field Production Operations." Oil & Gas Journal, 81 (June 20, 1983): 96-98.

1595. "China to Expand Yangbajain Geothermal Station." Oil & Gas Journal, 82 (Sept. 10, 1984): 96-97.

1596. Rahmer, B.A. "China Moves Toward Economic Growth." The Petroleum Economist, 46 (Jan. 1979): 13-15.

1597. "Special Report on China." Petroleum Economist, 48 (Nov. 1981): 745-98.

1598. Rahmer, B.A. "China: Industrial Plans on Target." Petroleum Economist, 50 (Jan. 1983): 12-14.

1599. Hough, G.V. "Energy Policy and Development." Petroleum Economist, 50 (Nov. 1983): 415+.

1600. Woodard, K. "China's Energy Prospects." [review article] Problems of Communism, 29 (Jan. 1980): 61-67.

1601. Odell, P.R. "World Energy in the 1980s: The Significance of Non-OPEC Oil Supplies." Scottish Journal of Political Economy, 26:3 (Nov. 1979): 215-31.

ECONOMIC GEOGRAPHY
[730]

1602. Pask, R. "Northeast China Is Transformed." Geographical Magazine, 53 (Oct. 1980): 53-60.

MANPOWER, LABOR, POPULATION
[800]

MANPOWER TRAINING AND ALLOCATION, LABOR FORCE AND SUPPLY
[810]

1603. "Not Just an Employment Agency: The Ziang Labour Service Companies." Beijing Review, 25 (Sept. 27, 1982): 23-26.

1604. "Computer Training of Cadres Planned." Beijing Review, 27 (March 26, 1984): 9-10.

1605. "Training Cadres on a Large Scale." Beijing Review, 27 (April 2, 1984): 24-31.

1606. "Longer Careers Urged for Women." Beijing Review, 27 (July 23, 1984): 31.

1607. "Rational Flow of Skilled Personnel." Beijing Review, 27 (Aug. 6, 1984): 4-5.

1608. "Labour Companies Offer More Jobs." Beijing Review, 28 (Sept. 9, 1985): 9-10.

1609. "Coolies for Hire." The Economist, 273 (Oct. 6, 1979): 83-84.

1610. "Excess into Asset." [unemployment] The Economist, 283 (June 12, 1982): 29-30.

1611. Sredl, H.J. "Industrial Training in China." Training and Development Journal, 34 (July/Aug. 1980): 52-54.

1612. Nadler, L. and Nadler, Z. "China: An HRD Study Tour." Training and Development Journal, 36 (March 1982): 50-51.

1613. Burns, S.K. "When Training Is the Law." *Training and Development Journal*, 38 (Oct. 1984): 29-31.

LABOR MARKETS, PUBLIC POLICY
[820]

1614. Xin, L. "China Promotes Innovative Jobs Policy." *Beijing Review*, 28 (Nov. 11, 1985): 4-5.

1615. "China: Any Reasonable Offer Considered." *The Economist*, 296 (Aug. 10, 1985): 30.

Theory of Labor Markets and Leisure
[821]

1616. "Communist Attitude Towards Labor." *Beijing Review*, 25 (Nov. 8, 1982): 3-4.

Public Policy, Role of Government
[822]

1617. "Reform of the Employment System." *Beijing Review*, 26 (April 4, 1983): 5-6.

1618. "Job-Hungry Young Have Beijing on Edge." *Business Week* (Feb. 16, 1981): 44.

1619. Shirk, S.L. "Recent Chinese Labour Policies and the Transformation of Industrial Organization in China." *The China Quarterly*, 88 (Dec. 1981): 575-93.

1620. Conroy, R. "New Domestic Policy." *The Far Eastern Economic Review*, 127 (March 28, 1985): 60-61.

Labor Mobility, National and International Migration
[823]

1621. "China: The Newest Export: Legions of Laborers." *Business Week* (Jan. 28, 1980): 59.

1622. Breeze, R. "Peking's People Exports." The Far Eastern Economic Review, 106 (Nov. 30, 1979): 68-69.

Labor Market Studies,
Wages, Employment
[824]

1623. "Employment Question." Beijing Review, 24 (May 25, 1981): 3-4.

1624. "Improving Wage System." Beijing Review, 24 (June 22, 1981): 5-6.

1625. "Employment Transformation." Beijing Review, 24 (Nov. 2, 1981): 25-28.

1626. "Decision on Solving Urban Employment Problem." Beijing Review, 25 (Feb. 8, 1982): 21-24.

1627. "Trend Toward Stable Urban Employment." Beijing Review, 25 (Sept. 27, 1982): 20-27.

1628. "46.5 Million People Employed Since 1977." Beijing Review, 26 (Sept. 19, 1983): 7-8.

1629. Korzec, M. and Whyte, M.K. "Reading Notes: The Chinese Wage System." The China Quarterly, 86 (June 1981): 248-73.

1630. He, X. "An Investigation into the Current Compensation System for Mental and Manual Labor." Chinese Economic Studies, 18:1 (Fall 1984): 77-95.

1631. "Job-Free." The Economist, 271 (June 30, 1979): 65-66.

1632. "Work Pays." The Economist, 273 (Dec. 29, 1979): 28-29.

1633. "Mass Unemployment." The Economist, 293 (Nov. 3, 1984): 74.

1634. "The High Price of Cheap Labor." The Economist, 295 (April 6, 1985): 70.

1635. Vinals, J.M. "Energy-Capital Substitution, Wage Flexibility, and Aggregate Output Supply." European Economics Review, 26 (Oct./Nov. 1984): 229-45.

1636. Breeze, R. "Battling the Bonus Blues." The Far Eastern Economic Review, 108 (June 6, 1980): 52.

1637. Bonavia, D. "Jobless Generation." The Far Eastern Economic Review, 111 (March 6-12, 1981): 30-31.

1638. Yue, G. "Employment, Wages and Social Security in China." International Labour Review, 124:4 (July/Aug. 1985): 411-22.

1639. Gittings, J. "Wages and Management in China." Journal of Contemporary Asia, 9:1 (1979): 53-66.

1640. Rawski, T.G. "Economic Growth and Employment in China." World Development, 7:8-9 (Aug./Sept 1979): 767-82.

1641. White, G. "Urban Employment and Labor Allocation Policies in Post-Mao China." World Development, 10 (Aug. 1982): 613-32.

Labor Productivity
[825]

1642. Wortzel, L.M. "Incentive Mechanism and Remuneration in China." [policies of the 11th Central Committee] Asian Survey, 21 (Sept. 1981): 961-76.

1643. Field, R.M. "Slow Growth of Labour Productivity in Chinese Industry (1952-81)." The China Quarterly, 96 (Dec. 1983): 641-64.

1644. Mailey, A.M. and Llobera, J.R. "K.A. Wittfogel and the Asiatic Mode of Production: A Reappraisal." Sociological Review, n.s. 27 (Aug. 1979): 541-59.

Labor Markets:
Demographic Characteristics
[826]

1645. "Population and Employment." Beijing Review, 26 (March 28, 1983): 17-22.

1646. "Occupation of the Employed Population." [3rd national census] Beijing Review, 27 (May 14, 1984): 24-25.

1647. Yu, G. "The Key Lies in Enhancing Economic Efficiency." Chinese Economic Studies, 17:4 (Summer 1984): 99-109.

1648. Zong, H. "Reduce the Consumption of Materialized Labor." Chinese Economic Studies, 18:4 (Summer 1985): 62-70.

1649. "Employing China's Millions." The Economist, 274 (Feb. 16, 1980): 108-9.

1650. Banister, J. "The Analysis of Recent Data on the Population of China." Population Development Review, 10:2 (June 1984): 241-71.

1651. White, G. "Urban Employment and Labour Allocation Policies in Post-Mao China." World Development, 10:8 (Aug. 1982): 613-32.

TRADE UNIONS, COLLECTIVE BARGAINING, LABOR-MANAGEMENT RELATIONS
[830]

1652. Tung, R.L. "Patterns of Motivation in Chinese Industrial Enterprises." The Academy of Management Review, 6 (July 1981): 481-89.

1653. "Provisional Regulations Concerning Congresses of Workers and Staff Members in State-Owned Industrial Enterprises." Beijing Review, 24 (Sept. 17, 1981): 16-19.

1654. "Worker's Movement Enters a New Period." Beijing Review, 27 (Feb. 13, 1984): 17-21.

1655. "Developments Over the Past Year." Beijing Review, 27 (May 14, 1984): 13-14.

1656. "Workers to Play Greater Role." Beijing Review, 28 (May 6, 1985): 7-8.

1657. Henley, J.S. and Chen, P.K.N. "A Note on the Appearance, Disappearance, and Re-Appearance of Dual Functioning Trade Unions in the People's Republic of China: Comment." British Journal of Industrial Relations, 19:1 (March 1981): 87-93.

1658. Yuan, Z. "On the Question of Understanding the Improvement in the Method of Bonus Distribution." Chinese Economic Studies, 18:4 (Summer 1985): 78-86.

1659. Lockett, M. "Self-Management in China?" Economic Analysis and Worker's Management, 15:1 (1981): 85-114.

1660. Loong, P. "Arguing for Arbitration." The Far Eastern Economic Review, 109 (Sept. 19, 1980): 144-45.

1661. Helburn, I.B. and Shearer, J.C. "Human Resources and Industrial Relations in China: A Time of Ferment." Industrial and Labor Relations Review, 38 (Oct. 1984): 3-15.

1662. Yang, C. "'Mass Line' Accounting in China." Management Accounting, 62 (May 1981): 13-17.

1663. Debats, K.E. "Impressions of China." [personnel management; panel discussion] Personnel Journal, 62 (July 1983): 520+.

DEMOGRAPHIC ECONOMICS
[840]

1664. Brown, L.R. "One Is Enough." Across the Board, 19 (March 1982): 27-28.

1665. "Controlling Population Growth." Beijing Review, 26 (Feb. 14, 1983): 21-28.

1666. "China Pushes Efforts to Reduce Population Growth." Chemical and Engineering News, 63 (Jan. 1985): 50-51.

1667. Reaves, J.A. "China's Reliance on Abortion a Concern." Chicago Tribune (June 16, 1985): sec. 5:5.

1668. Ng, P.P.T. "Planned Fertility and Fertility Socialization in Kwantung." The China Quarterly, 78 (June 1979): 351-59.

1669. Bianco, L. "Birth Control in China: Local Data and Their Reliability." The China Quarterly, 85 (March 1981): 119-37.

1670. Chen, C.H.C. and Tyler, C.W. "Demographic Implications of Family Size Alternatives in the People's Republic of China." The China Quarterly, 89 (March 1982): 65-73.

1671. Wong, S.-L. "Consequences of China's New Population Policy." The China Quarterly, 98 (June 1984): 220-40.

1672. Banister, J. "Population Policy and Trends in China, 1978-83." The China Quarterly, 100 (Dec. 1984): 717-41.

1673. Burstein, D. "Counting to a Billion." [computers used for census purposes] Datamation, 29 (March 1983): 183-85+.

1674. Beedham, B., et al. "One Is Best, Two Is Most." The Economist, 273 (Dec. 29, 1979): 24-25.

1675. Fraser, S.E. "One Is Fine, Two Is More Than Adequate." The Far Eastern Economic Review, 106 (Oct. 5, 1979): 61-62.

1676. Lyle, K.C. "China's Birth Planning: Organization Since the Cultural Revolution." Human Organization, 39 (Summer 1980): 197-201.

1677. Huang, L.J. "Planned Fertility of One-Couple/One-Child Policy in the People's Republic of China." Journal of Marriage and the Family, 44 (Aug. 1982): 775-84.

1678. Cheng Xiangmeng. "The One-Child Population Policy, Modernization, and the Extended Chinese Family." Journal of Marriage and the Family, 47 (Feb. 1985): 193-202.

1679. Platte, E. "China's Fertility Transition: The One-Child Campaign." Pacific Affairs, 57 (Winter 1984/85): 646-71.

1680. Pernia, E.M. "Some Impressions on China's Population and Demographic Research." Philippine Economic Journal, 19:3-4 (1980): 534-36.

1681. Coale, A.J. "Population Trends, Population Policy, and Population Studies in China." Population and Development Review, 7:1 (March 1981): 85-97.

1682. Chengrui, L. "On the Results of the Chinese Census." Population and Development Review, 9:2 (June 1983): 326-44.

1683. Oshima, H.T. "The Industrial and Demographic Transitions in East Asia." Population and Development Review, 9:4 (Dec. 1983): 583-607.

1684. Coale, A.J. "A Further Note on Chinese Population Statistics." Population and Development Review, 7:3 (Sept. 1981): 512-18.

1685. Birdsall, N. and Jamison, D.T. "Income and Other Factors Influencing Fertility in China." Population and Development Review, 9:4 (Dec. 1983): 651-75.

1686. Bongaarts, J. and Greenhalgh, A. "An Alternative to the One-Child Policy in China." Population and Development Review, 11:4 (Dec. 1985): 585-617.

1687. Zeng, Y. and Vaupel, I.W. "Marriage and Fertility in China: A Graphical Analysis." Population and Development Review, 11:4 (Dec. 1985): 721-36.

1688. Wolf, A.P. "The Preeminent Role of Government Intervention in China's Family Revolution." Population and Development Review, 12:1 (March 1986): 101-16.

1689. Miles, R.E., Jr. "Age of Discontinuity." Population Bulletin, 34 (Dec. 1979): 42-48.

1690. Yu, Y.C. "The Population Policy of China." Population Studies, 33:1 (March 1979): 125-42.

1691. Lavely, W.R. "The Rural Chinese Fertility Transition." Population Studies, 38:3 (Nov. 1984): 354-84.

1692. Wiltgen, R. and Herschede, F. "Marxism and Chinese Population Policies." Review of Radical Political Economies, 14:4 (Winter 1982): 18-22.

1693. Stacey, J. "Toward a Theory of Family and Revolution: Reflections on the Chinese Case." Social Problems, 26 (June 1979): 499-508.

1694. Mosher, S.W. "How China Uses U.N. Aid for Forced Abortions." The Wall Street Journal (May 13, 1985).

1695. Croll, E.J. "Production Versus Reproduction: A Threat to China's Developing Strategy." World Development, 11:6 (June 1983): 467-81.

HUMAN CAPITAL
[850]

1696. Tien, H.Y. "China: Demographic Billionaire." Population Bulletin, 38 (April 23, 1983): 2-42.

1697. Orazgel'dyev, M. "Training Cadres from the Local Population in the Central Asian Republics." Problems of Economics, 22:5 (Sept. 1979): 22-35.

WELFARE PROGRAMS, CONSUMER ECONOMICS, URBAN AND REGIONAL ECONOMICS
[900]

WELFARE, HEALTH, AND EDUCATION
[910]

1698. Lampton, D.M. "Roots of Interprovincial Inequality in Education and Health Services in China." _American Political Science Review_, 73 (June 1979): 459-77.

General Welfare Programs
[911]

1699. Vermeer, E.B. "Social Welfare Provisions and the Limits of Inequality in Contemporary China." _Asian Survey_, 19 (Sept. 1979): 856-80.

1700. Mok, B.H. "In the Service of Socialism: Social Welfare in China." _Social Work_, 28 (Aug. 1983): 269-72.

Education
[912]

1701. Seifman, E. "China's Key Schools: A New Educational Mandate." _Asian Affairs_, 10 (Feb. 1979): 42-50.

1702. Ogden, S. "China's Social Sciences: Prospects for Teaching and Research." _Asian Survey_, 22 (July 1982): 581-608.

1703. Thrope, S., et al. "[Americans] Teaching in China: What We Give, What We Get." _Asian Survey_, 23 (Nov. 1983): 1182-1208.

1704. "Spare-Time College for Beijing's Staff and Workers." _Beijing Review_, 24 (March 16, 1981): 31.

1705. "Special Feature/Youth." Beijing Review, 24 (July 27, 1981): 18-29.

1706. "College Graduates Assigned Work." Beijing Review, 25 (Sept. 13, 1982): 8-9.

1707. "New Skills for a New Era." Beijing Review, 25 (Oct. 18, 1982): 23-25.

1708. "Education in China: The Past Four Years." Beijing Review, 26 (Jan. 24, 1983): 22-30.

1709. "Higher Education." Beijing Review, 26 (Jan. 24, 1983): 28-30.

1710. "New Experiments in Colleges." Beijing Review, 26 (Feb. 21, 1983): 6-7.

1711. "After Careful Thought and Comparison--Increased Conviction in Marxism Among Beijing University Students." Beijing Review, 26 (May 2, 1983): 20-23.

1712. "Five-Year Plan for Higher Education." Beijing Review, 26 (June 6, 1983): 7-8.

1713. "Education for Modernization." Beijing Review, 26 (July 11, 1983): 4-5.

1714. "Employers Eager for Local Grads." Beijing Review, 27 (July 23, 1984): 7-8.

1715. Zhang Zeyu. "University Uses New Educational System." [Shenzhen University] Beijing Review, 28 (Feb. 11, 1985): 17-19.

1716. "Changing Educational Theory and Methods." Beijing Review, 28 (June 17, 1985): 19-21.

1717. Wu, N. "Students Find Alternate Paths to Success." Beijing Review, 28 (July 22, 1985): 22-24.

1718. "Officials to Train Provincial Teachers." Beijing Review, 28 (Aug. 5, 1985): 6-8.

1719. "[U.S.-Chinese] Educators Share Teaching Ideas." Beijing Review, 28 (Sept. 2, 1985): 34.

Welfare, Health, and Education

1720. "Qualified Teachers Urgently Needed." Beijing Review, 28 (Dec. 16, 1985): 9-10.

1721. Han, B. "Dalian Center Trains Entrepreneurs." Beijing Review, 28 (Dec. 30, 1985): 18-20.

1722. "A U.S.-Style B-School for Communist Managers." Business Week (Oct. 18, 1982): 82.

1723. Pepper, S. "Chinese Education After Mao: Two Steps Forward, Two Steps Back and Begin Again?" The China Quarterly, 81 (March 1980): 1-65.

1724. Munro, R. "Settling Accounts with the Cultural Revolution at Beijing University, 1977-78." The China Quarterly, 82 (June 1980): 308-33.

1725. Gold, T.B. "Back to the City: The Return of Shanghai's Educated Youth." The China Quarterly, 84 (Dec. 1980): 755-70.

1726. Kent, A. "Red and Expert: The Revolution in Education at Shanghai Teacher's University, 1975-76." The China Quarterly, 86 (June 1981): 304-21.

1727. Emerson, J.P. "Urban School-Leavers and Unemployment in China." The China Quarterly, 93 (March 1983): 1-16.

1728. Broaded, C.M. "Higher Education Policy Changes and Stratification in China." The China Quarterly, 93 (March 1983): 125-37.

1729. Hayhoe, R.E.S. "A Comparative Approach to the Cultural Dynamics of Sino-Western Educational Co-Operation." The China Quarterly, 104 (Dec. 1985): 676-99.

1730. Fang, I. "Report on the State of Science and Education." Chinese Economic Studies, 12:1-2 (Fall/Winter 1978/79): 124-38.

1731. "Yankee Traders Will Find Cathay Has Its Own MBAs." Christian Science Monitor, 76 (Nov. 5, 1984): 29.

1732. "From Red Guard to MBA Student: How to Get Ahead in China." Christian Science Monitor, 77 (June 18, 1985): 1.

Welfare Programs. Consumer/Urban & Regional Economics

1733. "Capitalist Tail Wags Chinese Dog." [education & manpower need] The Economist, 272 (Sept. 1, 1979): 71.

1734. "Ideology Finally Takes a Back Seat." The Far Eastern Economic Review, 103 (March 16, 1979): 54+.

1735. Chrysler, K.M. "Experimental University Put on Ice in Special Zone." The Far Eastern Economic Review, 131 (Feb. 6, 1986): 44-45.

1736. "Teaching Management to Marxists." Fortune, 103 (March 23, 1981): 102-3.

1737. "China Gropes for Perfect Blend of Management Techniques." International Management, 38 (April 1983): 57-58.

1738. Zhao Fu San. "Social Science and China's Modernization." International Social Science Journal, 34 (1982): 347-56.

1739. Zong, B. and Hildebrandt, H.W. "Business Communication in the People's Republic of China." [Beijing Institute of Foreign Trade] Journal of Business Communication, 20 (Winter 1983): 25-32.

1740. "China Turns to U.S. for Aid in Setting Up Business Schools (part 1)." Journal of Commerce and Commercial, 351 (Feb. 3, 1982): 1A.

1741. "China Should 'Flesh Out' Management Education Programs (part 2)." Journal of Commerce and Commercial, 351 (Feb. 4, 1982): 3A.

1742. Herschede, F. "Chinese Education and Economic Development: An Analysis of Mao Zedong's Contributions." Journal of Developing Areas, 14:4 (July 1980): 447-67.

1743. Jamison, D.T. "Child Malnutrition and School Performance in China." Journal of Development Economics, 20:2 (March 1986): 299-309.

1744. Lun, N.H. Ng. "Role of Hong Kong-Educated Chinese in the Shaping of Modern China." Modern Asian Studies, 17 (Feb. 1983): 137-63.

1745. Hartzell, B. "Why Study Chinese Personnel Management?" Personnel Journal, 61 (Oct. 1982): 724.

1746. Mangahas, M. "Notes on Economics in China." Philippine Economic Journal, 19:3-4 (1980): 526-33.

1747. "China's Society Biggest Challenge to Modernization." Research and Development, 27 (Feb. 1985): 128+.

1748. Pincus, F.L. "Higher Education and Socialist Transformation in the People's Republic of China Since 1970: A Critical Analysis." Review of Radical Political Economics, 11:1 (Spring 1979): 24-37.

1749. Gamson, Z.F. "China 1980: Whither the Revolution?" Social Policy, 10 (Jan. 1980): 31-35.

1750. "The Great Walls of China; American Teachers in China." Today's Education, 71 (Feb./March 1982): 50.

1751. Sredl, H.J. "Industrial Training in China." Training and Development Journal, 34 (July/Aug. 1980): 52-54.

1752. Nadler, L. and Nadler, Z. "China: An HRD Study Tour." Training and Development Journal, 36 (March 1982): 50-60.

1753. White, G. "Higher Education and Social Redistribution in a Socialist Society: The Chinese Case." World Development, 9 (Feb. 1981): 149-66.

Economics of Health
[913]

1754. Abrams, H.K. "Occupational Medicine in the People's Republic of China." Journal of Occupational Medicine, 22 (Aug. 1980): 553-57.

Economics of the Elderly
[914]

1755. "When Leaders or Professionals Retire." Beijing Review, 26 (May 9, 1983): 30-33.

1756. "Pension Predictions and Problems." Beijing Review, 27 (Oct. 22, 1984): 31-32.

1757. Treas, J. "Socialist Organization and Economic Development in China: Latent Consequences for the Aged." Gerontologist, 19 (Feb. 1979): 34-43.

1758. Cherry, R.L. and Magnusen-Martinson, S. "Modernization and the Status of the Aged in China: Decline or Equalization." Sociological Quarterly, 22 (Spring 1981): 253-61.

Economics of Crime
[915]

1759. "Crimebusters, China Style." Advertising Age, 55 (Oct. 8, 1984): 20.

1760. Zafanolli, W. "A Brief Outline of China's Second Economy." Asian Survey, 25 (July 1985): 715-36.

1761. "Crackdown on Economic Crimes." Beijing Review, 26 (Aug. 15, 1983): 7-8.

1762. "Stiffer Penalties for Black Money Mart." Beijing Review, 28 (May 13, 1985): 7-8.

1763. "Officials Sacked for Car Import Racket." Beijing Review, 28 (Aug. 12, 1985): 8-9.

1764. "Scandal Poses Another Threat to China's Economic Reform." [corruption among party cadres on Hainan Island] Christian Science Monitor, 77 (Aug. 6, 1985): 9.

1765. Lee, M. "Corruption: The Dark Side of the Liberalism Coin." The Far Eastern Economic Review, 127 (March 21, 1985): 68-69.

1766. Taylor, W.A. "Corruption Slows China's Turnabout." U.S. News & World Report (May 20, 1985): 53.

Minorities and Discrimination
[916]

1767. Lu, Y. "Minority People Living in the Capital." Beijing Review, 28 (July 15, 1985): 19-23.

1768. Delfs, R. "Economic Monitor; The Turning Point." The Far Eastern Economic Review, 127 (Jan. 24, 1985): 72.

CONSUMER ECONOMICS, LEVELS AND STANDARDS OF LIVING
[920]

1769. Burstein, D. "Consumers Offer Critique of Ads, Consumerism." Advertising Age, 52 (Dec. 14, 1981): sec. 2:S4.

1770. Whitney, R. "China in the Struggle for Economic Prosperity." Asian Affairs, 13 (Oct. 1982): 171-73.

1771. "More Consumer Goods." Beijing Review, 24 (April 20, 1981): 7-8.

1772. "Commercial Reforms: Smooth Circulation of Goods." Beijing Review, 24 (June 1, 1981): 20-29.

1773. "Individual Economy." Beijing Review, 24 (Aug. 17, 1981): 3-4.

1774. Li Chengrui and Zhang Zhongji. "Remarkable Improvement in Living Standards." Beijing Review, 25 (April 26, 1982): 15-18+.

1775. "State Subsidizes Living Costs." Beijing Review, 25 (Oct. 25, 1982): 6-7.

1776. Zhong He. "Standard of Living and Economic Construction." Beijing Review, 26 (Feb. 21, 1983): 12-13+.

1777. "Recent Changes in the Working Class." Beijing Review, 26 (June 6, 1983): 5-6.

1778. "Smooth Commodity Flow Needs More Avenues." Beijing Review, 27 (June 18, 1984): 27-30.

1779. "Surly Sales Clerks Dis-Serving People." Beijing Review, 28 (June 24, 1985): 9-10.

1780. Li, C. "Economic Reform Brings Better Life." Beijing Review, 28 (July 22, 1985): 15-22.

1781. "Nation Censures Shabby Products." Beijing Review, 28 (Sept. 23, 1985): 8-10.

1782. "Beijing Residents Seek Housemaids." Beijing Review, 28 (Nov. 25, 1985): 9-10.

1783. Nickum, J.E. and Schak, D.C. "Living Standards and Economic Development in Shanghai and Taiwan." The China Quarterly, 77 (March 1979): 25-49.

1784. Klatt, W. "Staff of Life: Living Standards in China, 1977-81." The China Quarterly, 93 (March 1983): 17-50.

1785. Lardy, N.R. "Consumption and Living Standards in China; 1978-83." The China Quarterly, 100 (Dec. 1984): 849-65.

1786. Shengming, Y. "Earnings, Prices, and Lives." Chinese Economic Studies, 17:2 (Winter 1983/84): 31-36.

1787. "Interest in Cosmetics Mounts in China." Drug and Cosmetic Industry, 135 (Nov. 1984): 52.

1788. "China Curbs Consumer Binge." Dun's Business Month, 126 (July 1985): 6-7.

1789. "China: Politics of the Purse." The Economist, 281 (Dec. 19, 1981): 43-44.

1790. "Grab a Quick Yuan While You Can." The Economist, 285 (Oct. 30-Nov. 5, 1982): 45-46.

1791. Bonavia, D. "China Turns to the Consumer." The Far Eastern Economic Review, 107 (March 7, 1980): 92.

1792. "Modernisation's Black Side." [black market] The Far Eastern Economic Review, 108 (April 4, 1980): 115-16.

Urban Economics

1793. Bonavia, D. "Concentrating--At Last--On the Quality of Life." The Far Eastern Economic Review, 113 (Sept. 25-Oct. 1, 1981): 45-46+.

1794. Thorelli, H.B. "First Survey of China's Middle-Class Consumers Finds 8% Own Refrigerators, But 75% Own TVs." Marketing News, 17 (Feb. 18, 1983): 16+.

1795. Smil, V. "Food Production and Quality of Diet in China." Population Development Review, 12:1 (March 1986): 25-45.

1796. Seigle, N.R. "Shopping Around in China." Stores, 62 (April 1980): 23-27.

1797. "China's Media Boom: Rapid Growth in TV Sets, Stations." Television Radio Age, 32 (May 13, 1985): 52.

1798. Martin, Robert P. "A Second Revolution Brings New Opportunities, New Challenges." U.S. News & World Report (June 24, 1985): 34-38.

1799. Bennett, A. "Peking Shoppers Stock Up on Foodstuffs Prior to Sweeping Price Increases Today." The Wall Street Journal (May 10, 1985).

1800. "Chinese Are Aflutter About U.S. Fast Food." The Wall Street Journal (Sept. 11, 1985).

URBAN ECONOMICS
[930]

1801. Ma, L.J.C. "Chinese Approach to City Planning: Policy, Administration, and Action." Asian Survey, 19 (Sept. 1979): 838-55.

1802. Orleans, L.A. and Burnham, L. "The Enigma of China's Urban Population." Asian Survey, 24 (July 1984): 788-804.

1803. Chang, S.D. "Modernization and China's Urban Development." Association of American Geographers, Annals, 71 (June 1981): 202-19.

1804. "More Urban Housing." Beijing Review, 24 (Jan. 12, 1981): 7-8.

1805. "Urban Income Increases." Beijing Review, 24 (Jan. 26, 1981): 6-7.

1806. "Smaller Cities Essential to Growth." Beijing Review, 25 (Nov. 8, 1982): 6-7.

1807. "Prosperity Follows Industrial Development." Beijing Review, 27 (May 28, 1984): 27-29.

1808. Fei Xiaotong. "Further Explorations of Small Towns." Beijing Review, 28 (April 8, 1985): 24-26; (April 29, 1985): 22-23.

1809. "Life Gets Better for City Dwellers." Beijing Review, 28 (July 8, 1985): 9-10.

1810. "'Horizontal' Links Key to Urban Reform." Beijing Review, 29 (April 7, 1986): 6-8.

1811. Kam, W.C. and Xueqiang, X. "Urban Population Growth and Urbanization in China Since 1949: Reconstructing a Baseline." The China Quarterly, 104 (Dec. 1985): 585-613.

1812. Jianzhong, T. and Ma, L.J.C. "Evolution of Urban Collective Enterprises in China." The China Quarterly, 104 (Dec. 1985): 614-40.

1813. Zhang, Z. "A Mirror for Urban Economic Reforms." Chinese Economic Studies, 19:2 (Winter 1985/86): 86-92.

1814. Mosher, S. "Lure of the Big City." The Far Eastern Economic Review, 110 (Oct. 31-Nov. 6, 1980): 12-13.

1815. Nolan, P. and White, G. "Urban Bias, Rural Bias or State Bias." Journal of Developmental Studies, 20:3 (April 1984): 52-81.

1816. Slater, G. "Chinese Approach to Housing Development Reveals Planning/Modernization Alternatives." Journal of Housing, 38 (May 1981): 256-61.

1817. Friedman, B.S. "Public Housing in China: Policies and Practices." Journal of Housing, 40 (May/June 1983): 82-85.

1818. Sawers, L. "Urban Planning in the Soviet Union and China." Monthly Review, 28 (March 1977): 34-48; 30 (June 1978): 58-64; (Dec. 1978): 62-63; 31 (Nov. 1979): 58-62.

1819. Hershkovitz, L. "The Fruits of Ambivalence: China's Urban Individual Economy." Pacific Affairs, 58 (Fall 1985): 427-50.

1820. Rowe, W.T. "Urban Policy in China." [review article] Problems of Communism, 33 (Nov./Dec. 1984): 75-80.

REGIONAL ECONOMICS
[940]

1821. Prybyla, J.S. "China's Special Economic Zones." ACES Bulletin, 26:4 (Winter 1984): 1-23.

1822. Weiss, J. "Shining Light on the People's Republic of China's SEZs (Special Economic Zones)." American Import/Export Management, 99 (Dec. 1983): 24-26.

1823. Stephani, G. "Special Economic Zones and Economic Policy in China." Annals of Public and Co-operative Economy, 54:3 (July-Sept. 1983): 289-312.

1824. Yeh, A.G.-O. and Xu, X. "Provincial Variation of Urbanization and Urban Primacy in China." Annals of Regional Science, 18:3 (Nov. 1984): 1-20.

1825. "Regions Co-Operate for Mutual Growth." Beijing Review, 27 (Oct. 22, 1984): 9-10.

1826. "Report from Shenzhen." Beijing Review, 27 (Nov. 26, 1984): 19-22.

1827. Wei Liming. "Shanghai Opens Its Arms to the World." Beijing Review, 28 (Jan. 14, 1985): 19-26.

1828. "Fourteen Open Cities Make Headway." Beijing Review, 28 (April 15, 1985): 8-9.

1829. "NPC Sidelights: Economic Reforms Succeed in Pilot Cities." Beijing Review, 28 (April 15, 1985): 19-21.

1830. Lu Yun. "Port City Profiles: Sagacious Tianjin Businessmen on the Go." Beijing Review, 28 (May 20, 1985): 20-24.

1831. "Landlocked Province Opens to the World." [Sichuan] Beijing Review, 28 (July 1, 1985): 20-24+.

1832. Zhang, Z. "South Fujian: Golden Delta for Investment." Beijing Review, 28 (Aug. 26, 1985): 25-28.

1833. "Cities Secure More Foreign Contracts." Beijing Review, 28 (Oct. 14, 1985): 29-30.

1834. Xia Zhen and Yue Haitao. "Changzhou's Urban Reform." Beijing Review, 29 (June 16, 1986): 4-20; (June 23, 1986): 24-27+; (June 30, 1986): 22-25; (July 21, 1986): 21-23.

1835. "Regional Exchange Draws Gains." Beijing Review, 29 (June 23, 1986): 6-8.

1836. "Heilongjiang Improves Foreign Trade." Beijing Review, 29 (Aug. 4, 1986): 28-29.

1837. "Heilongjiang Province Attracts Foreign Capital." Beijing Review, 29 (Aug. 18, 1986): 28-29.

1838. "How Trade Zones Are Luring Foreign Investors." Business Week (Jan. 11, 1982): 50-51.

1839. Chung, P. "Fully Utilize and Actively Develop Shanghai's Industry and Make Even Greater Contributions Toward the Realization of the Four Modernizations." Chinese Economic Studies, 12:1-2 (Fall/Winter 1978/79): 153-68.

1840. Cai, B. "Shanghai's Foreign Trade and Its Prospects." Chinese Economic Studies, 14:1 (Fall 1980): 79-93.

1841. Wang, M. and Chen, Y. "On the Nature of Asian Export Processing Zones and China's Special Economic Zones." Chinese Economic Studies, 19:2 (Winter 1985/86): 8-24.

1842. Shi, X. "Is the Economy of China's Special Economic Zones State Capitalist in Nature?" Chinese Economic Studies, 19:2 (Winter 1985/86): 25-40.

1843. Su, Y. "A Brief Discussion of the Economic Nature of China's Special Economic Zones." *Chinese Economic Studies*, 19:2 (Winter 1985/86): 41-58.

1844. Liang, X. "Shenzhen: Opening to the World." *Chinese Economic Studies*, 19:2 (Winter 1985/86): 73-78.

1845. Zou, E. "Special Economic Zone Typifies Open Policy." *Chinese Economic Studies*, 19:2 (Winter 1985/86): 79-85.

1846. Solinger, D.J. "China's New Economic Policies and the Local Industrial Political Process: The Case of Wuhan." *Comparative Politics*, 18 (July 1986): 379-99.

1847. Sit, V.F.S. "The Special Economic Zones of China: A New Type of Export Processing Zone?" *Developing Economies*, 32:1 (March 1985): 69-87.

1848. "The SEZs: Problems, Successes, Outlook." *East Asian Executive Reports*, 7 (July 15, 1985): 15-18.

1849. "China's Enterprise Zones: Leaky Capitalist Enclaves." *The Economist*, 285 (Nov. 27, 1982): 87-88.

1850. Yeh, G.-O.A. "Development of the Special Economic Zones in Shenzhen, the People's Republic of China." *Ekistics*, 52 (March/April 1985): 154-61.

1851. Ma, T. "Turning Full Circle." [Dalian] *The Far Eastern Economic Review*, 126 (Oct. 11, 1984): 97-98.

1852. Kraar, L. "A Little Touch of Capitalism." [special economic zones] *Fortune*, 107 (April 18, 1983): 120-22+.

1853. Phillips, D.R. "Special Economic Zones in China's Modernization: Changing Policies and Changing Fortunes." *National Westminster Bank Quarterly Review* (Feb. 1986): 37-49.

1854. Oborne, M.W. "China's Early Windows on the World: The Special Economic Zones." *The OECD Observer*, 133 (March 1985): 11-12+.

1855. Mills, W. deB. "Leadership Change in China's Provinces." *Problems of Communism*, 34 (May/June 1985): 24-40.

1856. Williamson, J. "Telecommunications Expansion in China's Economic Zones." Telephony, 208 (April 22, 1985): 61+.

1857. Taylor, W. "China's Lures for Western Know-How." [Xiamen] U.S. News & World Report (Dec. 10, 1984).

1858. "China Cuts Back SEZs But Says Long Term Policy Isn't Changed." The Wall Street Journal (Aug. 6, 1985).

SOCIAL PROBLEMS AND CONDITIONS
[950]

1859. Myers, J.T. "China--The 'Germs' of Modernization." Asian Survey, 35 (Oct. 1985): 981-97.

1860. Schell, O. "Return of China's Curbside Capitalists." Asia, 3 (July 1980): 10-11+.

1861. "Small Families on the Rise." Beijing Review, 24 (May 4, 1981): 23-25.

1862. Zhou Yan. "On China's Current Class Struggle." Beijing Review, 25 (Aug. 16, 1982): 17-19.

1863. "Chinese Workers' Spare-Time Activities." Beijing Review, 26 (Oct. 3, 1983): 23-25.

1864. "The Lives of Working Women in China." Beijing Review, 27 (Sept. 10, 1984): 28-32.

1865. "Safeguarding Women's Rights." Beijing Review, 27 (Oct. 15, 1984): 27-29.

1866. "Women Reject Return to Home." Beijing Review, 28 (Feb. 4, 1985): 9-10.

1867. "Women Claiming Their Legal Rights." Beijing Review, 28 (March 11, 1985): 9-10.

1868. "Open Policy Must Consider Culture." Beijing Review, 29 (March 10, 1986): 26-27.

1869. Kwong, J. "Is Everyone Equal Before the System of Grades? Social Background and Opportunities in

Social Problems and Conditions 161

China." British Journal of Sociology, 34 (March 1983): 93-108.

1870. Ellithorpe, H. and Young, L.H. "Wave of Dissidence Sweeps China's Youth." Business Week (July 9, 1979): 39.

1871. Jones, D.E., et al. "To Get Rich Is Glorious--But Not Too Rich." Business Week (April 15, 1985): 60-61.

1872. Reaves, J.A. "In Peking, Getting Cab No Hack Job." Chicago Tribune (May 19, 1985): sec. 1:3.

1873. Robinson, J.C. "Of Women and Washing Machines: Employment, Housework, and the Reproduction of Motherhood in Socialist China." The China Quarterly, 101 (March 1985): 32-57.

1874. Gold, T.B. "After Comradeship: Personal Relations in China Since the Cultural Revolution." The China Quarterly, 104 (Dec. 1985): 657-75.

1875. Sheng, R. "Outsiders' Perception of the Chinese." The Columbia Journal of World Business, 14 (Summer 1979): 16-22.

1876. Nee, V. "Political and Social Bases of China's Four Modernizations." The Columbia Journal of World Business, 14 (Summer 1979): 23-32.

1877. Nakamura, J.I. and Miyamoto, M. "Social Structure and Population Change: A Comparative Study of Tokugawa, Japan and Ching, China." Economic Development and Cultural Change, 30 (Jan. 1982): 229-69.

1878. "China's New Pin-Striped Heros." [management] The Economist, 292 (Sept. 8, 1984): 76.

1879. "Learning to Love the Money-Makers." The Economist, 293 (Dec. 22, 1984): 76.

1880. Davies, D. "China: Rebirth of the Individual." The Far Eastern Economic Review, 105 (Aug. 17, 1979): 51-56.

1881. Bonavia, D. "Foreigners Are Dangerous--Official." The Far Eastern Economic Review, 118 (Oct. 1-7, 1982): 45-46.

Welfare Programs, Consumer/Urban & Regional Economics

1882. Bonavia, D. "Real Problem May Be in the Provinces." The Far Eastern Economic Review, 122 (Nov. 3, 1983): 44-45.

1883. Wang Gungwu. "The Cult of Progress and Heroes of Today." The Far Eastern Economic Review, 129 (Aug. 8, 1985): 32-33.

1884. Mandle, J.R. "Strategies of Change in Paternalistic Socialism: The Case of China." International Journal of Social Economics, 11:3-4 (1984): 3-11.

1885. Nevis, E.C. "Using an American Perspective in Understanding Another Culture: Toward a Hierarchy of Needs for the People's Republic of China." The Journal of Applied Behavioral Science, 19 (1983): 249-64.

1886. Lindsay, C.P. and Dempsey, B.L. "Ten Painfully Learned Lessons About Working in China: The Insights of Two American Behavioral Scientists." The Journal of Applied Behavioral Science, 19 (1983): 265-76.

1887. Robinson, D.C. "Changing Functions of Mass Media in the People's Republic of China." Journal of Communication, 31 (Autumn 1981): 58-73.

1888. Lee, H.Y. "The Implications of Reform for Ideology, State and Society in China." Journal of International Affairs, 39 (Winter 1986): 77-89.

1889. Watson, R.S. "Class Differences and Affinal Relations in South China." Man, 16 (Dec. 1981): 593-615.

1890. Thurston, A.F. "Victims of China's Cultural Revolution: The Invisible Wounds." Pacific Affairs, 57 (Winter 1984/85): 599-620; (Spring 1985): 5-27.

1891. Lampton, D.M. "New Revolution in China's Social Policy." Problems of Communism, 28 (Dec. 1979): 16-33.

1892. Dreyer, J.T. "Limits of the Permissible in China." Problems of Communism, 29 (Nov. 1980): 48-65.

1893. Rosen, S. "Prosperity, Privatization, and China's Youth." Problems of Communism, 34 (March/April 1985): 1-28.

1894. Bridgwater, C.A. "The Chinese Middle Class." *Psychology Today*, 18 (Jan. 1984): 76-77.

1895. Dalsimer, M. and Nisonoff, L. "The New Economic Readjustment Policies: Implications for Chinese Urban Working Women." *Review of Radical Political Economics*, 16:1 (Spring 1984): 17-43.

1896. Bennett, Amanda. "In Today's China, Road to Romance Begins at the Bank." *The Wall Street Journal* (Oct. 4, 1985).

1897. "Chinese Trains Supply Chopsticks, But Bring Your Own Washcloth." *The Wall Street Journal* (Oct. 17, 1985).

1898. Chan, A. "Images of China's Social Structure." *World Politics*, 34 (April 1982): 295-323.

AUTHOR INDEX

Aaron, B. 585
Abrams, H.K. 1754
Adkins, L. 698
Agres, T. 284, 1206
Ahmad, V. 75
Alford, E.P. 631
Allen, B. 738
Altmann, H. 741
Altschul, J. 908
An, B. 794
An Zhiguo 353
Anand, V. 571
Anderson, M.H. 1145
Argall, G.O., Jr. 1529, 1530
Arguelles, R. 568
Asher, J. 640
Au-Yeung, P.K. 479
Aznam, S. 844

Baark, E. 1179
Bachman, D. 9
Balassa, B. 204
Banister, J. 1650, 1672
Barbour, N.S. 1209
Barty, E. 701
Baum, R. 105
Beedham, B. 165, 1099, 1674
Beng, P. 221
Bennett, A. 78, 136, 366, 408, 1070, 1071, 1799
Bennett, Amanda 138, 633, 752, 1896
Bennett, G. 321
Bennett, W.R. 738
Berges, A. 463
Bergson, A. 241
Bernstein, P.W. 720
Bhalla, A.S. 1201
Bianco, L. 1669
Bin, L. 1594

Birdsall, N. 1685
Blecher, H. 1324
Blecher, M. 293, 1425, 1441, 1442
Bleiberg, R.M. 891
Block, P.M. 1035
Bohn, J. 1022
Bonavia, D. 121, 122, 123, 169, 174, 175, 257, 258, 262, 265, 359, 441, 500, 563, 815, 822, 830, 918, 921, 922, 923, 989, 1100, 1147, 1202, 1203, 1225, 1243, 1274, 1277, 1318, 1368, 1426, 1427, 1428, 1578, 1637, 1791, 1793, 1881, 1882
Bongaarts, J. 1686
Boreham, G.F. 426
Borkenau, F. 791
Bowring, P. 995
Braddick, B. 1108
Brauchli, M. 481
Bray, D. 763
Breeze, R. 173, 373, 1622, 1636
Breithaupt, H. 818
Brick, P. 766
Bridgwater, C.A. 1894
Broaded, C.M. 1728
Brody, M. 648
Bronfenbrenner, M. 144
Brookins, C. 240
Brown, B. 987
Brown, L.R. 1664
Bu, M. 893
Bucknall, K.B. 1090
Burnham, L. 1802
Burns, J.P. 1389
Burns, S.K. 1613

Burstein, D. 1127, 1673, 1769
Buzo, A. 845
Byrd, W. 242, 1162, 1163

Cai, B. 1840
Cai, T. 616
Callahan, J.M. 597
Chan, A. 1435, 1898
Chan, T.S. 695
Chan, W.K.K. 1152
Chanda, N. 564, 714, 1468
Chang, H. 458
Chang, H.Y. 1177
Chang, S.D. 1803
Chao, K. 538
Chase, D. 1125, 1131
Chastain, C.E. 1107
Chen, C.H.C. 1670
Chen, N.-R. 99, 372, 529, 530, 606, 609, 637, 660, 665, 673, 875, 959
Chen, P.K.N. 1657
Chen, Y. 402, 1841
Chen Muhua 522
Chen Qiwei 485
Cheng, C. 129
Cheng, E. 180, 928
Cheng, J.Y.S. 858
Cheng, L. 1531
Cheng, P.C. 388
Cheng, Y. 490
Cheng-Fang, Y. 4
Cheng Xiangmeng 1678
Chengrui, L. 1682
Chengwu, M. 1594
Chenn, D.L. 1380
Cherry, R.L. 1758
Ch'iang, L. 489
Chihren, C. Lin 231
Chin, R.Q.P. 285
Chinn, D.L. 392, 1244, 1254, 1372, 1374
Ch'iu-li, Y. 1300
Chong, P. 260
Chongwei, J. 617
Chow, G.C. 74, 390

Chrysler, K.M. 1735
Chu, L. 627
Chu, W.W. 1229
Chun Yun 355
Chung, P. 1839
Chyba, C.F. 644
Clad, J. 833, 836, 837
Clutterbuck, D. 184
Coale, A.J. 1681, 1684
Cohen, J.A. 708
Cohen, S.E. 1124
Conroy, R. 1195, 1620
Conway, M. 1161
Coplin, W.D. 132
Crane, A.T. 548
Cremer, J. 1373
Croll, E.J. 1695
Cromley, R.G. 1538
Crook, F.W. 17, 1379
Crow, P. 629
Curry, L. 825, 1130

Dahlby, T. 909, 912, 916
Dai, Y. 54, 472
Dalsimer, M. 1895
Das, D.K. 739
Davies, D. 170, 1880
Davis, D.A. 1223
Dean, B.V. 728
Debats, K.E. 1663
De Boer, A.J. 1346
Delfs, R. 116, 117, 118, 119, 120, 121, 124, 125, 126, 176, 263, 264, 266, 267, 268, 269, 271, 312, 322, 360, 361, 362, 374, 387, 406, 432, 443, 446, 460, 462, 509, 623, 711, 828, 830, 849, 924, 993, 1075, 1101, 1155, 1165, 1204, 1227, 1253, 1275, 1276, 1344, 1370, 1429, 1430, 1431, 1474, 1580, 1581, 1583, 1584, 1768
Dempsey, B.L. 1886
Deng, H. 56
Denis, R. 584

Author Index

Dennis, R.D. 106
Denny, D.L. 666, 1066
Dernberger, R.F. 376, 1234
de Wulf, L. 272, 447
Deyan, Z. 535
DiFederico, E.M. 667
Ding, N. 306
Dirksen, E. 424, 511
Disney, R. 1228
Djurovic, M. 1571
do Rosario, L. 178, 363, 565, 566, 567, 715, 841, 843, 881, 1119, 1319
Donath, B. 574, 1143
Dong, F. 235
Dong Dasheng 37
Dong Fureng 576
Dreyer, J.T. 1892
Du, R. 1384, 1424
Du Runsheng 1394, 1408
Dunphy, J.F. 687

Edmond, M. 1489, 1490
Ehrlich, P. 1044
Elliott, D. 787
Ellithorpe, H. 1870
Ellman, M. 279
Emerson, J.D. 1492
Emerson, J.P. 1727
Emerson, R.S. 853
Emmott, B. 809
Enright, R.J. 1502
Erlich, P. 620
Etienne, G. 1326
Ewing, H.G. 722

Fan, C. 1294
Fan, L.-S. 1294
Fang, I. 1730
Fang, S. 964
Fei Xiaotong 1418, 1808
Feng, J. 1423
Fewsmith, J. 1390
Field, R.M. 1222, 1643
Fischer, W.A. 1111
Fodella, G. 313, 377
Forger, G. 744

Frank, A.D. 1587
Frank, R. 988
Fraser, S.E. 1675
Fridley, D. 1521
Friedman, B.S. 1817
Friedman, E. 1296, 1325
Friedman, J.A. 134
Fu, Z. 794
Fung, Vigor 137
Fureng, D. 158
Furst, A. 1043

Gaeta, J.J. 938
Gage, T.J. 1129
Galting, J. 246
Gamson, Z.F. 1749
Garb, F.A. 1535
Garcia-Borras, T. 1029
Gasper, D.R. 197
Ge, Z. 416
Geng, Y. 347
Gengmo, L. 369
Gibson, W.D. 1031
Gigot, P. 430, 919
Gilbert, N. 1052
Gittings, J. 1104, 1639
Gold, T.B. 1725, 1874
Goldman, C. 852
Goldman, M.I. 863
Goldstein, C. 851
Gong, Z. 379
Goodman, D.S.G. 133
Goodstadt, L. 429, 559, 813, 879, 907
Gorman, T.D. 856
Gray, P. 728
Greenhalgh, A. 1686
Griffin, K. 1434, 1437
Grives, R.T. 788
Grow, R.F. 796
Gu, N. 1196
Gu, S. 59, 60
Gullo, D.T. 526
Guo, H. 1171

Hake, B. 1233
Halpern, N.P. 1150

Ham Guang 1116
Hammer, A. 1002
Han, B. 951, 1217, 1721
Han, K. 877
Han, X. 735
Hanham, R.Q. 1177
Hansen, J. 857
Hanson, R. 992
Harrison, M. 1232
Hart, P. 1063
Hartzell, B. 1110, 1745
Hataye, J. 812
Hayhoe, R.E.S. 1729
He, J. 209, 1114
He, X. 63, 496, 1630
He Jianzhang 35, 1301
Helburn, I.B. 1661
Hendryx, S.R. 573
Henley, J.S. 1657
Herschede, F. 1488, 1692, 1742
Hershkovitz, L. 1819
Hester, S.B. 678
Hickok, S.A. 568
Hidasi, G. 141, 367
Hildebrandt, H.W. 1739
Hinkle, D.E. 678
Hinton, W.H. 1376
Holland, D.S. 1536
Hong, D. 65
Hong Yuangpeng 28
Hou, C.M. 372
Hough, G.V. 1599
Howard, F. 79
Howe, C. 614, 792
Hsu, R. 1283
Hsu, R.C. 1257
Hu Ruiliang 18
Hu Sheng 89
Huan, G. 1001
Huang, H. 382
Huang, L.J. 1677
Huang Pinfu 1306
Huang Zhengi 31
Huixiang, Z. 577
Hung-Ying, C. 1200

Ichimura, S. 532
Ignatius, A. 1475
Imai, H. 448
Ishikawa, S. 82, 296, 304
Ishimine, T. 502, 591
Islam, S. 835
Israel, E.A. 1069

Jacobs, P. 864
Jamison, D.T. 1685, 1743
Jao, J.C. 469
Jao, Y.C. 827
Jenkins, D. 817
Ji, C. 892, 962
Ji, X. 67
Ji, Z. 57
Ji Chongwei 25
Ji Jincheng 1303
Ji Zhengzhi 39
Jia, K. 64
Jia, L. 61
Jian, C. 349, 950
Jiang, Q. 62
Jianzhong, T. 1812
Jin, Q. 1192
Jin, R. 427
Jing Wei 953
Jiyun, T. 1164
Johnson, C. 297
Johnson, D.G. 1079, 1282
Johnson, G. 1371
Johnson, T.M. 1521
Jones, D. 610
Jones, D.E. 229, 788, 1871
Jones, G.C.L. 1511

Kadhim, M. 1488
Kallgren, J. 1364
Kam, W.C. 1811
Kambara, T. 1559
Kaye, L. 838
Keidel, A. 155, 876
Kemenade, W. van 821
Kent, A. 1726
Kharbanda, O.F. 1572
Kharbanda, V.P. 1573

Kilpatrick, J.A. 515, 1237
Kim, Shee Poon 764
King, R., Jr. 642, 643
Klatt, W. 101, 193, 375, 381, 638, 1784
Knaak, R. 194
Knight, B. 190
Knowles, H.A. 596
Kokubun, R. 793
Kolbenschlag, M. 569
Kornai, J. 480
Korzec, M. 1629
Kosenko, R. 575
Kraar, L. 162, 276, 516, 570, 719, 854, 855, 1053, 1852
Krafft, W. 897
Kraus, R. 1166
Kravis, I.B. 389
Kueh, Y.Y. 232, 517, 614, 1255, 1322, 1386
Kulkarni, V.G. 838, 1483
Kurata, P. 983
Kwong, J. 1869

Laaksonen, O. 1088
Lampton, D.M. 1698, 1891
Langston, N. 713, 925, 926, 990, 1481, 1482, 1484, 1485
Langston, Nancy 994
Lanier, A.R. 870
Lappen, A.A. 1148
Lardy, N.R. 42, 183, 1236, 1375, 1785
Lasdon, L.S. 1102
Lasuriat, G. 1466
Lau, E. 880
Lauriat, G. 1172, 1173, 1467, 1469
Lavely, W.R. 1691
Lee, D. 814, 816
Lee, H.Y. 1888
Lee, J. 529
Lee, J.L. 606, 608, 665, 673

Lee, M. 123, 445, 461, 820, 823, 826, 832, 834, 927, 980, 984, 985, 991, 996, 1141, 1472, 1484, 1579, 1765
Lee, P.N.-S. 103
Lee, R.W. 639
Leeming, F. 1347
Lewis, J. 822, 982
Li, C. 1780
Li, H. 498
Li, J. 306
Li, Z. 49
Li Chengrui 1774
Li Erhuang 1269
Li Haibo 1167
Li Honglin 484
Li Yongzhen 1334
Liang, X. 317, 1844
Liang Wensen 27
Lieberthal, K. 42, 127, 198, 275, 1149
Lin, L. 795
Lin, W. 61, 149
Lin, Z. 69, 963
Lin Jiyang 1305
Lindsay, C.P. 1886
Ling, X.Q. 280
Lipschutz, N. 743
Littler, C.R. 1112
Liu, A.P.L. 142
Liu, G. 44, 415, 873
Liu, H. 437, 602
Liu, K. 41
Liu, M. 115, 256, 259, 440, 560, 911, 914, 915, 975, 978, 1142, 1154, 1159, 1160, 1316, 1317, 1467
Liu, Y. 782
Liu, Z. 1249
Liu Guoguang 34, 401
Liu Mingfu 32
Liu Peiti 1305
Liu Sunhao 1308
Llobera, J.R. 1644
Lockett, M. 1112, 1659
Lomax, D.F. 1491

Loong, P. 261, 431, 459, 474, 477, 979, 986, 1087, 1471, 1660
Lowe, J.Y. 628
Lowe, L. 1109
Lu, B. 1340
Lu, L. 1249
Lu, Y. 1767
Lu Dong 1190
Lu Shijian 1313
Lu Yun 1357, 1358, 1359, 1412, 1830
Lucyk, C.L. 609
Lukin, V. 645
Lun, N.H. Ng 1744
Luo, G. 383
Lyle, K.C. 1676
Lyons, T.P. 185, 1175, 1176

Ma, H. 50, 305
Ma, L.J.C. 1801, 1812
Ma, T. 562, 711, 831, 1476, 1477, 1478, 1479, 1480, 1586, 1851
Ma Chuandong 1365
Ma Jiaju 36
McCall, J.B. 582
McClenahen, S. 725
McConnell, J.J. 463
McFarlane, B. 139, 294, 325, 422
McGaughey, H.W. 140
Maffry, A., Jr. 662
Magnusen-Martinson, S. 1758
Maher, P. 679
Maidment, P. 806
Maier, J.H. 1208
Mailey, A.M. 1644
Maitan, L. 245
Malenbaum, W. 308, 310, 311
Mandle, J.R. 1884
Mangahas, M. 6, 1746
Manion, M. 1078
Mann, P. 598, 647
Manning, R. 564, 712, 1582
Marer, P. 531
Mark, J. 1525

Martellaro, J.A. 745
Martin, Robert P. 1798
Matheson, J. 1020
Matsuda, M. 1128
Maxwell, B. 1048
Maxwell, N. 325, 1270, 1440
Meilach, Dona Z. 630
Meisner, M. 1302, 1425
Meng Lian 38
Merry, R.W. 753
Miles, R.E., Jr. 1689
Miller, W.H. 277, 723, 724, 725, 726, 727
Mills, W. deB. 1855
Mirsky, J. 131, 187
Miyamoto, M. 1877
Modic, S.J. 278
Mok, B.H. 1700
Monk, L.B. 530, 677
Moore, T.E. 1019
Mosher, S. 1814
Mosher, S.W. 1694
Mozingo, D. 536
Mu Jiajun 1303
Mullins, P.E. 1017
Mun, K.C. 695
Mundheim, R. 605
Munro, R. 1724
Munson, S. 584
Munthe-Kaas, H. 1466
Muqiao, X. 45
Myers, H.E. 239
Myers, J.T. 1859
Myers, R.H. 1265

Nadler, L. 1612, 1752
Nadler, Z. 1612, 1752
Nakamura, J.I. 1877
Namiotkiewicz, W. 634
Nations, R. 624, 840, 1585
Nee, V. 104, 1876
Nehemkis, A. 1027
Nehemkis, P. 1027
Nelson, J.F. 935
Nevans, A. 997
Nevis, E.C. 746, 1885
Ng, P.P.T. 1668

Author Index

Nickum, J.E. 1783
Nisonoff, L. 1895
Nolan, P. 295, 1293, 1815
Norton, R.E. 721
Noumoff, S.J. 790

Oborne, M.W. 1854
Odell, P.R. 1601
Ogden, S. 1702
Okita, S. 717
Oksenberg, M. 102, 718
O'Leary, G. 1369, 1436
O'Leary, M.K. 132
Orazgel'dyev, M. 1697
Orleans, L.A. 163, 1802
Oshima, H.T. 1683
Oster, M. 965

Paine, S. 323
Pairault, T. 81, 191
Pang, C.M. 1346
Pannell, C.W. 1278
Parker, M.A. 1507
Pask, R. 1602
Passow, S. 1132
Pastore, S.L. 890
Pearlstine, N. 716
Peebles, G. 3
Peng, T. 53
Pepper, S. 1723
Perkins, D.H. 303
Pernia, E.M. 1680
Perry, E.J. 1387
Peter, L.J. 1231
Petr, J.L. 734
Phanachet, U. 577
Phen Zhen 225
Phillips, C.H. 107
Phillips, D.R. 1486, 1853
Pincus, F.L. 1748
Platte, E. 1679
Pollack, J.D. 867
Prybyla, J.S. 8, 85, 156, 199, 201, 324, 1821
Putterman, L. 1245, 1323
Pye, L.W. 572

Quihua, Q. 16, 539
Qureshi, M.A. 1573

Rada, E.L. 1247
Rae, A.E.I. 159, 961
Rahmer, B.A. 1596, 1598
Raichur, S. 10
Ram, M. 561
Ranis, G. 157, 488, 507
Rawski, T.G. 5, 384, 1640
Reaves, J.A. 1667, 1872
Redding, S.G. 1105
Reeder, J.A. 1085, 1242
Reid, B.C. 1146
Renpu, W. 1508
Reynolds, B.L. 196, 503, 1295
Richman, B. 682
Riskin, C. 1439
Rivers, C. 1473
Robinson, D.C. 1887
Robinson, J.C. 1873
Robinson, T.W. 742
Roby, J.L. 1054
Rong, Y. 1003
Roscoe, B. 850
Rosen, S. 1893
Ross, L. 243
Ross, M.C. 693
Ross, R.S. 757
Rowan, R. 182
Rowe, W.T. 1820
Rowley, A. 442, 444, 474, 621, 913, 917, 929, 982, 1470
Ruggles, R.L., Jr. 690

Saith, A. 1419, 1434
Samli, C. 575
Sandeman, H. 799, 800
Saunders, I. 909
Sawer, M. 76
Sawers, L. 1818
Schak, D.C. 1783
Scheibla, S.H. 649
Schell, O. 1860
Schlarbaum, G.G. 463

Schram, S.R.   2
Schroeder, P.E.   688
Schwartz, L.   703, 1224
Scott, R.W.   1532
Scott, W.E.   1463
Scouton, W.   668
Segal, G.   1097
Segal, J.   1519
Seidman, H.L.   607
Seifman, E.   1701
Seigle, N.R.   1796
Seixas, S.   740
Selden, M.   80, 1328
Seltzer, R.J.   230
Shanweng, W.   1501
Shao, M.   506
Shearer, J.C.   1661
Shen, L.   44
Shen, P.   890
Sheng, R.   1875
Shengming, Y.   48, 1786
Shi, J.   691
Shi, X.   1842
Shi Qian   1366
Shi Wen   1544
Shi Yan   1452
Shih, A.   479
Shirk, S.L.   1619
Shoazhi, S.   164
Sicular, T.   1320
Siggins, M.   789
Simon, D.F.   287, 646
Sit, V.   171
Sit, V.F.S.   1847
Skinner, G.W.   1251, 1339
Slater, G.   1816
Smil, V.   709, 1537, 1568, 1569, 1570, 1795
Smith, C.   842, 846, 847, 848
Smith, D.C.   1073
Smith, P.   824
Smock, D.   1524
Snyderman, N.   556
Solinger, D.J.   200, 326, 590, 1156, 1846
Song, D.   1314

Song Ping   343, 354
Spielmann, P.   751
Sredl, H.J.   1611, 1751
Srodes, J.   475, 704, 706, 981
Stacey, J.   1693
Stavis, B.   1280, 1302, 1329
Stefani, G.   425
Stephani, G.   1823
Sterba, James P.   754
Stoltenberg, C.D.   694
Stone, B.   1268
Stuart, D.T.   540
Su, Y.   1843
Subhan, M.   819
Sucheki, S.M.   748, 749, 1230
Suharchuk, G.D.   504
Sun, Y.   412
Sun Lingen   1312
Sun Ping   451
Sun Ru   26
Sun Yousheng   1312
Sundrum, R.M.   309
Surls, F.M.   514, 1256, 1298
Suttmeier, R.P.   548
Szuprowicz, B.O.   583

Tan, K.C.   1432
Tang, A.M.   1281, 1367
Tang, H.   876
Tanzer, A.   273, 824, 1051
Tarpey, J.P.   1026
Taylor, W.   1857
Taylor, W.A.   1766
Terrill, R.   7, 181
Tharp, M.   1226
Thaxton, R.   70
Thomas, P.G.   1008
Thompson, D.N.   154
Thorelli, H.B.   1794
Thrope, S.   1703
Thurston, A.F.   1890
Tian, J.   98, 400, 403
Tian, Y.   1080
Tian Jiasen   1365
Tian Jiyun   396

Author Index

Tian Yun 1395
Tien, H.Y. 1696
Timmer, C.P. 1264
Ting Yi-Lan 1588
Tow, W.T. 540, 1205
Toy, C.D. 1018
Travers, L. 1297
Travers, S.L. 378
Treas, J. 1757
Trescott, P.B. 192
Trewhitt, J. 687
Tsumuri, Y. 545, 689, 1039
Tung, R.L. 547, 737, 1089, 1652
Tuthill, M. 1060
Tyler, C.W. 1670

Unger, J. 1435
Unna, W. 1121

Valigra, L. 1058
Vander Weele, R. 1106
Vaupel, I.W. 1687
Vermeer, E.B. 1266, 1420, 1699
Vertzberger, Y. 765
Verzariu, P. 153, 528
Vinals, J.M. 1577, 1635
Volti, R. 135

Walder, A.G. 1194
Walker, K.R. 1252, 1267
Walter, C.E. 439
Wan Dianwu 40
Wang, A. 1218
Wang, B. 428
Wang, D. 601, 1168, 1262
Wang, D.A. 413
Wang, G.C. 541
Wang, H. 320
Wang, J. 492, 499
Wang, K. 497
Wang, L. 493
Wang, M. 1841
Wang, N.T. 692
Wang, R. 316
Wang, W. 65

Wang Bingquian 386
Wang Dacheng 467
Wang Gungwu 1883
Wang Haibo 23
Wang Yongnian 1309
Wang Zhenzhi 25
Wang Ziade 1308
Warrington, B. 865
Warrington, M.B. 582
Wassermann, U. 501, 581
Watson, A. 1369, 1378, 1383, 1388, 1436
Watson, R.S. 1889
Watzmann, A. 1035
Webber, D. 683
Wei, L. 1117
Wei, Y. 779
Wei Liming 1827
Wei Yuming 942
Weiss, J. 1822
Weiwen, Z. 33
Westlake, M. 710, 1174
Wethington, O. 674
White, G. 1641, 1651, 1753, 1815
White, L.T. 73, 186
Whitney, R. 1770
Whyte, M.K. 1629
Wiedemann, K.M. 143
Wiens, T.B. 1345
Wiesemeyer, J. 549
Williams, T. 705
Williamson, J. 1856
Wilson, D. 86, 205
Wilson, Dick 206
Wilson, P. 622
Wiltgen, R. 1692
Wisniewski, M. 829
Wolf, A.P. 1688
Wong, S.-L. 1083, 1671
Woodard, K. 1523, 1600
Woodward, D. 1377
Woodward, P. 672
Woodward, P.L. 676
Wortzel, L.M. 1642
Wu, F. 1007
Wu, F.T. 878
Wu, F.W. 1005

Wu, J. 68, 160, 306
Wu, N. 1717
Wu Jinglian 23
Wu Kaitai 1301
Wyllie, R.J.M. 756, 1528

Xia Zhen 1834
Xiang Qiyuan 31
Xiao, X. 655
Xin, L. 1614
Xiong Yi 1304
Xu, D. 202, 371, 411
Xu, R. 471
Xu, S. 494
Xu, X. 1824
Xu Dixin 20
Xue, M. 11, 13, 414, 1444
Xueqiang, X. 1811
Xuezeng, L. 48

Yan, X. 519
Yan Kalin 450
Yang, C. 1662
Yang, D.J. 506, 1193
Yang, Y. 60
Yang Ruichun 1310
Yang Ye 59
Yao, Y. 328
Ye Yinsong 1093
Yee, H.S. 872
Yefang, S. 46
Yeh, A.G.-O. 1824, 1850
Yeh, K.C. 233
Yian, Y. 427
Yonekawa, S. 83
You Lin 21
Young, L.H. 533, 1870
Yu, G. 55, 357, 1084, 1094, 1647
Yu, M. 1534
Yu, Q. 379, 1574
Yu, X. 963
Yu, Y.C. 1690
Yu Bing 1543
Yu Guangyuan 24, 29, 30, 47, 1241
Yuan, W. 492

Yuan, Z. 427, 1340, 1658,
Yuan Zhenmin 603
Yuanzheng, L. 518
Yue, G. 1638
Yue, P. 52
Yue, W. 43
Yue Haitao 785, 1834
Yung-kuei, C. 1299

Zafanolli, W. 1760
Zagoria, D.S. 274
Zaiyi, T. 1506
Zang Danan 1312
Zeng, Y. 1687
Zhang, E. 434
Zhang, L. 495
Zhang, Z. 51, 635, 1314, 1813, 1832
Zhang Chaozun 31
Zhang Jinfu 1287
Zhang Liufang 1311
Zhang Wenxiao 19
Zhang Zeyu 438, 1715
Zhang Zhongji 299, 318, 1774
Zhao, B. 894
Zhao, L. 66
Zhao, Y. 487, 615
Zhao, Z. 12, 345, 356
Zhao Fu San 1738
Zhao Renwei 34
Zhao Ziyang 90, 219, 333, 351, 483, 510, 1291
Zheng Lizhi 1310
Zhifang, L. 1505
Zhong He 1776
Zhou, J. 207
Zhou, S. 84, 1197
Zhou, Z. 1550, 1551
Zhou Shu 1
Zhou Shulian 23
Zhou Yan 1862
Zhou Zhengdu 1307
Zhu, C. 410
Zhu Qingfang 300
Zhuang, Q. 657
Zi, Z. 657

Zong, B.   1739
Zong, H.   1648
Zou, E.   1845
Zou, S.   491

Zuo, M.   58
Zuo Xu   22
Zuoxiang, A.   1503
Zweig, D.   1348, 1422